**Springer Theses**

Recognizing Outstanding Ph.D. Research

## Aims and Scope

The series "Springer Theses" brings together a selection of the very best Ph.D. theses from around the world and across the physical sciences. Nominated and endorsed by two recognized specialists, each published volume has been selected for its scientific excellence and the high impact of its contents for the pertinent field of research. For greater accessibility to non-specialists, the published versions include an extended introduction, as well as a foreword by the student's supervisor explaining the special relevance of the work for the field. As a whole, the series will provide a valuable resource both for newcomers to the research fields described, and for other scientists seeking detailed background information on special questions. Finally, it provides an accredited documentation of the valuable contributions made by today's younger generation of scientists.

## Theses are accepted into the series by invited nomination only and must fulfill all of the following criteria

- They must be written in good English.
- The topic should fall within the confines of Chemistry, Physics, Earth Sciences, Engineering and related interdisciplinary fields such as Materials, Nanoscience, Chemical Engineering, Complex Systems and Biophysics.
- The work reported in the thesis must represent a significant scientific advance.
- If the thesis includes previously published material, permission to reproduce this must be gained from the respective copyright holder.
- They must have been examined and passed during the 12 months prior to nomination.
- Each thesis should include a foreword by the supervisor outlining the significance of its content.
- The theses should have a clearly defined structure including an introduction accessible to scientists not expert in that particular field.

More information about this series at http://www.springer.com/series/8790

Nan Guo

# Infrared Photodetectors Based on Low-Dimensional Materials

Doctoral Thesis accepted by the
University of Chinese Academy of Sciences, Beijing, China

Springer

*Author*
Dr. Nan Guo
Qian Xuesen Laboratory of Space Technology
China Academy of Space Technology
Beijing, China

*Supervisor*
Prof. Weida Hu
Shanghai Institute of Technical Physics
Chinese Academy of Sciences
Shanghai, China

ISSN 2190-5053 ISSN 2190-5061 (electronic)
Springer Theses
ISBN 978-981-13-2837-4 ISBN 978-981-13-2838-1 (eBook)
https://doi.org/10.1007/978-981-13-2838-1

Library of Congress Control Number: 2018957625



This Springer imprint is published by the registered company Springer Nature Singapore Pte Ltd.
The registered company address is: 152 Beach Road, #21-01/04 Gateway East, Singapore 189721, Singapore

# Supervisor's Foreword

In 2010, Dr. Nan Guo joined our group and began to study infrared photodetectors based on low-dimensional materials. Infrared photodetectors, which turn the invisible infrared radiant energy into easily measurable electric signals, are widely used in many fields, such as spectroscopy, night vision and object recognition. However, its development still faces many problems. The traditional response mechanisms cannot provide better device performance, which requires new materials and physical mechanisms. In recent years, the emergence of low-dimensional materials with unique properties, such as high carrier mobility, tunable bandgap, easy combination between different materials without lattice mismatch, provides new prospects to enhancing the performance of infrared photodetectors.

Dr. Nan Guo has made systemtic studies of infrared photodetectors through simulations and experiments. He analysed the response physical mechanism of low-dimensional materials including graphene, nanowires and transition metal dichalcogenides and proposed design schemes on the improvement of the device performance. Scanning photocurrent microscopy was used to investigate the photothermoelectric effect in graphene device. And, an active infrared imaging was performed using the graphene device as a point-like detector at room temperature. In order to solve the problem of weak light absorption in InAs nanowires, a core–shell-like structure was designed to realize photogating effect, thus achieving high-gain photodetection. A hybrid $WSe_2$-$In_2O_3$ nanowire structure was designed to solve the trade-off between the dark current and the photoconductive gain. We hope that the results presented in this thesis will provide useful insights into future researches of high-performance infrared photodetectors.

Shanghai, China
August 2018

Prof. Weida Hu

# Acknowledgements

Writing a thesis takes a lot of effort and time, but no worthwhile achievements could have been made without help from many people. The completion of this work is a good moment to perform the duty of returning thanks to them.

First of all, I would like to express my gratitude to my supervisor, Prof. Weida Hu, who has given me support and encouragement in scientific research during the 5 years of studies. I was deeply impressed by Prof. Hu's rigorous attitude to study and his deep understanding of the frontier of the subject. From the initial design of the detector structure, theoretical simulation and calculation, to the preparation of devices, conducting tests and data analysis, the results obtained all include the hard work of Prof. Hu. His strict guidance and requirements will make a lifelong benefit for me.

I sincerely thank Prof. Xuechu Shen, Prof. Wei Lu and Prof. Xiaoshuang Chen for their guidance and help in the analysis of experimental data and the revision of the paper. My thanks also go to Prof. Lei Liao and his team members in Wuhan University. Professor Liao has a wealth of experience in the fabrication of nanodevices. From him, I learned a lot of experimental skills. I also owe much to the teachers of the state key laboratory of infrared physics for their help and guidance in my scientific research.

Last but not least, I want to thank my parents for their love, care and encouragement, which have proven to be invaluable emotional support for me to move forward in pursuing both scientific discoveries and career advancement.

Beijing, China
August 2018

Dr. Nan Guo

# Contents


**1 Introduction** . . . . . . . . 1
1.1 Two-Dimensional Material-Based Photodetectors . . . . . . . . 1
1.2 Nanowire Photodetectors . . . . . . . . 3
References . . . . . . . . 5

**2 Infrared Imaging with Graphene Photodetectors** . . . . . . . . 7
2.1 Introduction . . . . . . . . 7
2.2 Results and Discussions . . . . . . . . 8
2.2.1 Device Structure and Photoresponse Measurement . . . . . . . . 8
2.2.2 Physical Mechanism . . . . . . . . 10
2.2.3 $V_{gs}$ Dependence of Photoresponse . . . . . . . . 12
2.2.4 Multispectral Active Infrared Imaging . . . . . . . . 15
2.3 Conclusions . . . . . . . . 17
2.4 Detailed Methods . . . . . . . . 18
References . . . . . . . . 18

**3 Highly-Efficient InAs Nanowire Phototransistors** . . . . . . . . 21
3.1 Introduction . . . . . . . . 21
3.2 Results and Discussions . . . . . . . . 22
3.2.1 Nanowire Growth and Device Fabrication . . . . . . . . 22
3.2.2 Physical Mechanism . . . . . . . . 24
3.2.3 Photoresponse Performance . . . . . . . . 24
3.3 Conclusions . . . . . . . . 32
References . . . . . . . . 32

**4 Hybrid $WSe_2$-$In_2O_3$ Phototransistors with Ultrahigh Detectivity** . . . 35
4.1 Introduction . . . . . . . . 35
4.2 Results and Discussions . . . . . . . . 36
4.2.1 Device Structure and Working Principle . . . . . . . . 36
4.2.2 Photoresponse Performance . . . . . . . . 38

4.2.3 Temporal Response Characterization . . . 42
4.2.4 Physical Mechanism . . . 44
4.3 Conclusions . . . 48
4.4 Detailed Methods . . . 49
References . . . 49

**5 Summary and Outlook** . . . 53

# Chapter 1
# Introduction

**Abstract** The aim of this chapter is to introduce photodetectors based on low-dimensional materials. It will start with an overview of the development of two-dimensional materials and their application in photodetection. Subsequently, the development of nanowire photodetectors and the obstacle in how to realize high gain in photodetection will be discussed.

**Keywords** Two-dimensional material · Nanowire · Infrared photodetector · Gain

## 1.1 Two-Dimensional Material-Based Photodetectors

The discovery of graphene, which is formed by single-layer carbon atoms with honeycomb lattice structure and is stable in natural environment [1], has aroused sustained enthusiasm for the exploration of a wide-ranging new two-dimensional (2D) materials since 2004. Numerous studies about the preparation, characterization and functional devices based on graphene have been conducted. Due to its unique characteristics of broadband photon absorption [2], ultrafast carrier dynamics [3] and high mobility [4], graphene is regarded as a highly promising material for photodetectors. However, graphene also suffers from several major obstacles in its application in this field. The absorptance of light in single-layer graphene is only 2.3% [5], while, high absorptance is often required for high performance photodetectors. Additionally, its zero bandgap leads to a large dark current and the fast carrier recombination processes in graphene also limit its responsivity. In recent years, more and more 2D materials such as transition-metal dichalcogenides (TMDs) [6–8], hexagonal boron nitride (h-BN) [9, 10], black phosphorus [11] and $PtSe_2$ [12] were explored. Due to the weak interaction between layers bonded by van der Waals forces, 2D materials can be obtained conveniently by mechanical exfoliation from their bulk forms.

After graphene, TMDs such as $MoS_2$, $MoTe_2$, $WS_2$, and $WSe_2$, have been studied extensively. These 2D semiconductor materials share one similarity in their energy band structures: monolayer materials have a direct bandgap, and multilayer materials have an indirect bandgap [13]. Moreover, as the layer number decreases, the bandgap



N. Guo, *Infrared Photodetectors Based on Low-Dimensional Materials*,
Springer Theses, https://doi.org/10.1007/978-981-13-2838-1_1

become wider and can vary over a large range (1–2 eV) [13], corresponding to the spectral range from visible to near infrared. In recent years, the newly discovered black phosphorus (BP) shows a direct bandgap of ~2 eV in monolayer and ~0.3 eV in its bulk form [11, 14]. The narrow bandgap extends the spectrum to mid infrared. Therefore, the appearance of 2D BP provides a new candidate for the research of mid-infrared photodetectors [15]. Overall, the thin and atomically flat interface of 2D materials are ideal to realize diverse heterostructures without lattice mismatch, and the combination of various 2D materials in one device can make up for each other's deficiencies to achieve novel physical mechanisms and improve device performance.

So far, many kinds of heterostructure photodetectors based on 2D materials have been reported. Kim et al. reported graphene-$MoS_2$-$WSe_2$-graphene vertical heterostructure with a photovoltaic effect [16]. The stack of n-type $MoS_2$ and p-type $WSe_2$ forms a vertical p-n junction. Graphene was used as electrodes directly on the top and at the bottom of the junction. The high mobility of graphene results in a fast carrier collection. Koppens et al. reported graphene-$WSe_2$-graphene vertical heterostructure encapsulated in h-BN [17]. In this sandwich structure, the $WSe_2$ flak acted as photoactive material, and monolayer graphene, placed below and above the $WSe_2$ flak, was used as source and drain electrodes to collect the carriers photoexcited in $WSe_2$. A response time of 5.5 ps comparable to that of graphene was obtained. Zhong et al. reported graphene/$Ta_2O_5$/graphene photodetectors [18]. Hot carrier tunnelling in this graphene double-layer heterostructure leads to a mid-infrared (3.2 μm) responsivity of 1.1 A $W^{-1}$ at room temperature.

These achievements have benefitted from the progress of photoelectric examination technologies, especially from the scanning photocurrent microscopy (SPM), an efficient approach in investigating the physical mechanism of photoresponse. In this measurement, an incident laser is focused onto the device through an objective lens using a confocal system. The photocurrents are recorded with the laser scanning the device surface, which is realized through a piezo-driven sample stage or a galvanometer mirror scanning system, and finally a photocurrent 2D map of the device can be obtained. SPM accurately reveals the generating position of photocurrents, contributing to the understanding of local optoelectronic characteristics of the device. In addition, by applying bias voltage to the device and studying the change of local photocurrent, more can be known about its energy band distribution and response physical mechanism.

In 2008, for the first time, Lee et al. used SPM to investigate the local photocurrents in graphene devices and reveal the photovoltaic effect at the contacts [19]. After that, SPM was widely adopted to study the photoresponse mechanisms of 2D materials based photodetectors. For example, in 2010, McEuen et al. reported photothermoelectric effect at graphene single-bilayer interface junction [20]. In 2011, Herrero et al. presented hot carrier-dominated intrinsic photoresponse in graphene p-n junction rather than photovoltaic effect [21]. In 2013, Duan et al. reported vertical graphene-$MoS_2$-graphene heterostructures with a high external quantum efficiency, and SPM was used to analyze the changes in the amplitude and polarity of photocurrents with gate modulation [22]. SPM has been proved as an ideal, visualized approach to unravel local changes in electronic structure of low-dimensional materials. In this thesis, Chap. 2 introduces and analyzes the photoresponse mechanism of graphene transistors using SPM, combined with Raman mapping and numerical simulation. And, an active infrared imaging is carried out by the as-fabricated graphene photodetectors.

## 1.2 Nanowire Photodetectors

Semiconductor nanowires (NWs) have attracted major attention due to their potential application in nanoelectronics and nanophotonics, and a variety of NWs have been synthesized and fabricated into photodetectors, including III-V compound semiconductor NWs (InP [23], InAs [24], GaAs [25], InGaAs [26], GaN [27]), IV semiconductor NWs (Si [28], Ge [29]), VI semiconductor NWs (Se [30], Te [31]), II-VI compound semiconductor NWs (CdTe [32], CdSe [33], ZnSe [34], CdS [35]) and metal oxide NWs (ZnO [36], $SnO_2$ [37], $V_2O_5$ [38]). Among them, InAs NW has attracted particular attention due to fact that the surface Fermi level pinning in the conduction band edge can result in an electron surface accumulation layer and simple formation of low resistance ohmic contacts [39, 40]. With InAs NW, its diameter of 50 and 80 nm leads to the mobility of 1500 and 3000 $cm^2/Vs$ [41, 42], respectively. The high mobility at room temperature makes it a good candidate for high frequency electronic devices. Han et al. studied the electric properties of InAs NW field effect transistors [43]. The $Al_2O_3$ passivation on NW surface reduces the density of surface defects and carrier recombination, improving the NW mobility. And, the hysteresis in transfer curve was reduced as well. In addition, InAs NWs have potential applications in gas sensing due to their rich surface states and electron surface accumulation layers. Du et al. reported the use of InAs NW back-gated transistor for gas sensing [44]. The results show that $H_2O$, isopropanol and alcohol can lead to a positive shift of the threshold voltage due to carrier transfer from NW to adsorbed gas molecules through surface defects. Offermans et al. used the InAs NW arrays that were grown vertically on substrate for $NO_2$ gas sensing by taking advantage of the same mechanism [45].

Despite the wide interest in the electronic devices and gas sensors made of InAs NW, insufficient study has been made of its photoresponse, which is mainly due to a large dark current arising from high background electron concentration. The suppression of dark current usually requires a high gate voltage, which, greatly raises the technical threshold of other components in high-performance photodetectors. Besides, the surface defect states limit the photoresponse. Hu et al. reported the single crystal InAs NWs grown by molecular beam epitaxy and fabricated them into back-gated transistors for photodetection [46]. In their study, the response was greatly affected by ambient gas molecules and could be improved by surface passivation. Gas molecules adsorbed on the surface of InAs NWs not only reduced the strength of built-in field at the NW-metal junction by electrostatic interaction, but also trapped the photocarriers with the defect centers on NW surface. Moreover, the responsivity was not high enough due to the small photosensitive area and low quantum efficiency.

Liu et al. reported InAs NWs synthesized by chemical vapor deposition (CVD) and then fabricated into back-gated transistors, demonstrating a wide spectrum detection ranging from 300 nm to 1100 nm, thanks to the NW narrow bandgap of ~0.354 eV [47]. Wei et al. reported an n-type InAs NWs which grew vertically on p-Si substrates [24]. The photodetection was realized through the heterojunction formed between InAs NWs and p-Si substrates. The quantum efficiency of the device without bias was ~20%. Therefore, though InAs NWs device has achieved wide spectrum detection, it is difficult to improve its responsivity due to the lack of a gain mechanism.

In 2007, Soci et al. reported ZnO NW ultraviolet high-gain photodetectors [36]. The operation mechanism was: under dark conditions, oxygen molecules adsorbed on the NW could capture free electrons and turn into oxygen ions [$O_2 + e^- \rightarrow O_2^-$]. After illumination, electron-hole pairs were generated in ZnO NW, and the holes would migrate to the surface due to band bending at NW surface, and then neutralize oxygen ions with negative charges [$h^+ + O_2^- \rightarrow O_2$], leading to the desorption of oxygen at NW surface. The unpaired electrons collected by electrodes would contribute to photocurrent, until the oxygen molecules were adsorbed on the NW again to capture electrons to be ionized. The internal conductive gain was up to $10^8$. By efficiently separating electron-hole pairs, the lifetime of photocarriers was greatly extended, and the reduction of the recombination achieved a high gain. Despite the high performance of ZnO NW, its responsivity is completely suppressed in the absence of oxygen. Moreover, the wide bandgap of ZnO NW makes it effective only for ultraviolet detection, which limits its application in wide spectrum detection.

In recent years, a promising approach based on photogating effect has been widely adopted to achieve high-gain photodetectors, in which the strongly light absorbing materials (sensitizers) were used to decorate the conductive channels, modulating the channel conductance. In 2012, Konstantatos et al. reported hybrid phototransistors that combined the colloidal quantum dots (CQDs) with graphene channel [48]. The built-in field formed at the graphene-CQDs interface separated the oppositely charged carriers photoexcited in CQDs. The trapped carriers in CQDs resulted in a photogate to enhance the conductance of graphene through capacitive coupling. The fabricated devices showed an ultrahigh photoconductive gain of ~$10^8$. After that, many CQDs-based phototransistors with photogating effect have been explored such as $MoS_2$-CQDs [49], InGaZnO-CQDs [50] and $WSe_2$-CQDs [51]. Unfortunately, this method which deposits sensitizers onto conductive channels is not applicable to NWs due to their small surface area.

How to apply this photogating effect to one-dimensional (1D) InAs NWs to achieve high gain is the starting point of our research. InAs NW with core/shell-like structure is introduced in Chap. 3 of this thesis. Photocarriers captured by NW shell modulate the channel conductance, leading to a photogating effect. An anomalous photocurrent with high gain at room temperature was obtained. Moreover, in Chap. 4, a novel hybrid phototransistor is introduced which consists of 2D $WSe_2$ and 1D $In_2O_3$ NW with ultrahigh detectivity based on photogating effect.

## References


1. K.S. Novoselov, A.K. Geim, S.V. Morozov, D. Jiang, Y. Zhang, S.V. Dubonos, I.V. Grigorieva, A.A. Firsov, Science **306**, 666 (2004)
2. J.M. Dawlaty, S. Shivaraman, J. Strait, P. George, M. Chandrashekhar, F. Rana, M.G. Spencer, D. Veksler, Y. Chen, Appl. Phys. Lett. **93**, 131905 (2008)
3. J.M. Dawlaty, S. Shivaraman, M. Chandrashekhar, F. Rana, M.G. Spencer, Appl. Phys. Lett. **92**, 042116 (2008)
4. K.I. Bolotin, K.J. Sikes, Z. Jiang, M. Klima, G. Fudenberg, J. Hone, P. Kim, H.L. Stormer, Solid State Commun. **146**, 351 (2008)
5. R.R. Nair, P. Blake, A.N. Grigorenko, K.S. Novoselov, T.J. Booth, T. Stauber, N.M.R. Peres, A.K. Geim, Science **320**, 1308 (2008)
6. B. Radisavljevic, A. Radenovic, J. Brivio, V. Giacometti, A. Kis, Nat. Nanotechnol. **6**, 147 (2011)
7. Z. Yin, H. Li, H. Li, L. Jiang, Y. Shi, Y. Sun, G. Lu, Q. Zhang, X. Chen, H. Zhang, ACS Nano **6**, 74 (2012)
8. W. Zhao, Z. Ghorannevis, L. Chu, M. Toh, C. Kloc, P.-H. Tan, G. Eda, ACS Nano **7**, 791 (2013)
9. K. Watanabe, T. Taniguchi, H. Kanda, Nat. Mater. **3**, 404 (2004)
10. L. Song, L. Ci, H. Lu, P.B. Sorokin, C. Jin, J. Ni, A.G. Kvashnin, D.G. Kvashnin, J. Lou, B.I. Yakobson, P.M. Ajayan, Nano Lett. **10**, 3209 (2010)
11. L. Li, Y. Yu, G.J. Ye, Q. Ge, X. Ou, H. Wu, D. Feng, X.H. Chen, Y. Zhang, Nat. Nanotechnol. **9**, 372 (2014)
12. Y. Zhao, J. Qiao, Z. Yu, P. Yu, K. Xu, S.P. Lau, W. Zhou, Z. Liu, X. Wang, W. Ji, Y. Chai, Adv. Mater. **29**, 1604230 (2017)
13. F. Xia, H. Wang, D. Xiao, M. Dubey, A. Ramasubramaniam, Nat. Photonics **8**, 899 (2014)
14. V. Tran, R. Soklaski, Y. Liang, L. Yang, Phys. Rev. B **89**, 235319 (2014)
15. M. Amani, E. Regan, J. Bullock, G.H. Ahn, A. Javey, ACS Nano **11**, 11724 (2017)
16. C.-H. Lee, G.-H. Lee, A.M. van der Zande, W. Chen, Y. Li, M. Han, X. Cui, G. Arefe, C. Nuckolls, T.F. Heinz, J. Guo, J. Hone, P. Kim, Nat. Nanotechnol. **9**, 676 (2014)
17. M. Massicotte, P. Schmidt, F. Vialla, K.G. Schädler, A. Reserbat-Plantey, K. Watanabe, T. Taniguchi, K.J. Tielrooij, F.H.L. Koppens, Nat. Nanotechnol. **11**, 42 (2016)
18. C.-H. Liu, Y.-C. Chang, T.B. Norris, Z. Zhong, Nat. Nanotechnol. **9**, 273 (2014)
19. E.J.H. Lee, K. Balasubramanian, R.T. Weitz, M. Burghard, K. Kern, Nat. Nanotechnol. **3**, 486 (2008)
20. X. Xu, N.M. Gabor, J.S. Alden, A.M. van der Zande, P.L. McEuen, Nano Lett. **10**, 562 (2010)
21. N.M. Gabor, J.C.W. Song, Q. Ma, N.L. Nair, T. Taychatanapat, K. Watanabe, T. Taniguchi, L.S. Levitov, P. Jarillo-Herrero, Science **334**, 648 (2011)
22. W.J. Yu, Y. Liu, H. Zhou, A. Yin, Z. Li, Y. Huang, X. Duan, Nat. Nanotechnol. **8**, 952 (2013)
23. J. Wang, M.S. Gudiksen, X. Duan, Y. Cui, C.M. Lieber, Science **293**, 1455 (2001)
24. W. Wei, X.-Y. Bao, C. Soci, Y. Ding, Z.-L. Wang, D. Wang, Nano Lett. **9**, 2926 (2009)
25. S. Thunich, L. Prechtel, D. Spirkoska, G. Abstreiter, A. Fontcuberta i Morral, A.W. Holleitner, Appl. Phys. Lett. **95**, 083111 (2009)
26. D.G. Cooke, F.A. Hegmann, Y.I. Mazur, Z.M. Wang, W. Black, H. Wen, G.J. Salamo, T.D. Mishima, G.D. Lian, M.B. Johnson, J. Appl. Phys. **103**, 023710 (2008)
27. M. Kang, J.-S. Lee, S.-K. Sim, H. Kim, B. Min, K. Cho, G.-T. Kim, M.-Y. Sung, S. Kim, H.S. Han, Jpn. J. Appl. Phys. **43**, 6868 (2004)
28. A. Zhang, H. Kim, J. Cheng, Y.-H. Lo, Nano Lett. **10**, 2117 (2010)
29. Y.H. Ahn, J. Park, Appl. Phys. Lett. **91**, 162102 (2007)
30. B. Gates, B. Mayers, B. Cattle, Y. Xia, Adv. Funct. Mater. **12**, 219 (2002)
31. Y. Wang, Z. Tang, P. Podsiadlo, Y. Elkasabi, J. Lahann, N.A. Kotov, Adv. Mater. **18**, 518 (2006)
32. Y. Yu, V. Protasenko, D. Jena, H. (Grace) Xing, M. Kuno, Nano Lett. **8**, 1352 (2008)
33. S.-C. Kung, W.E. van der Veer, F. Yang, K.C. Donavan, R.M. Penner, Nano Lett. **10**, 1481 (2010)

34. J. Salfi, U. Philipose, C.F. de Sousa, S. Aouba, H.E. Ruda, Appl. Phys. Lett. **89**, 261112 (2006)
35. Q. Li, R.M. Penner, Nano Lett. **5**, 1720 (2005)
36. C. Soci, A. Zhang, B. Xiang, S.A. Dayeh, D.P.R. Aplin, J. Park, X.Y. Bao, Y.H. Lo, D. Wang, Nano Lett. **7**, 1003 (2007)
37. S. Mathur, S. Barth, H. Shen, J.-C. Pyun, U. Werner, Small **1**, 713 (2005)
38. Z. Li, Z. Hu, J. Peng, C. Wu, Y. Yang, F. Feng, P. Gao, J. Yang, Y. Xie, Adv. Funct. Mater. **24**, 1821 (2014)
39. M. Noguchi, K. Hirakawa, T. Ikoma, Phys. Rev. Lett. **66**, 2243 (1991)
40. J.M. Woodall, J.L. Freeouf, G.D. Pettit, T. Jackson, P. Kirchner, J. Vac. Sci. Technol. **19**, 626 (1981)
41. E. Lind, A.I. Persson, L. Samuelson, L.-E. Wernersson, Nano Lett. **6**, 1842 (2006)
42. T. Bryllert, L.E. Wernersson, L.E. Froberg, L. Samuelson, IEEE Electron Device Lett. **27**, 323 (2006)
43. N. Han, F. Wang, J.J. Hou, S.P. Yip, H. Lin, F. Xiu, M. Fang, Z. Yang, X. Shi, G. Dong, T.F. Hung, J.C. Ho, Adv. Mater. **25**, 4445 (2013)
44. J. Du, D. Liang, H. Tang, X.P.A. Gao, Nano Lett. **9**, 4348 (2009)
45. P. Offermans, M. Crego-Calama, S.H. Brongersma, Nano Lett. **10**, 2412 (2010)
46. J. Miao, W. Hu, N. Guo, Z. Lu, X. Zou, L. Liao, S. Shi, P. Chen, Z. Fan, J.C. Ho, T.-X. Li, X.S. Chen, W. Lu, ACS Nano **8**, 3628 (2014)
47. Z. Liu, T. Luo, B. Liang, G. Chen, G. Yu, X. Xie, D. Chen, G. Shen, Nano Res. **6**, 775 (2013)
48. G. Konstantatos, M. Badioli, L. Gaudreau, J. Osmond, M. Bernechea, F.P.G. de Arquer, F. Gatti, F.H.L. Koppens, Nat. Nanotechnol. **7**, 363 (2012)
49. D. Kufer, I. Nikitskiy, T. Lasanta, G. Navickaite, F.H.L. Koppens, G. Konstantatos, Adv. Mater. **27**, 176 (2015)
50. D.K. Hwang, Y.T. Lee, H.S. Lee, Y.J. Lee, S.H. Shokouh, J. Kyhm, J. Lee, H.H. Kim, T.-H. Yoo, S.H. Nam, D.I. Son, B.-K. Ju, M.-C. Park, J.D. Song, W.K. Choi, S. Im, NPG Asia Mater. **8**, e233 (2016)
51. C. Hu, D. Dong, X. Yang, K. Qiao, D. Yang, H. Deng, S. Yuan, J. Khan, Y. Lan, H. Song, J. Tang, Adv. Funct. Mater. **27**, 1603605 (2017)

# Chapter 2
# Infrared Imaging with Graphene Photodetectors

**Abstract** This chapter will introduce the infrared photoresponse in graphene transistors. Photocurrents induced by photothermoelectric effect will be investigated using scanning photocurrent microscopy combined with Raman mapping and numerical simulation. A self-developed multispectral active infrared imaging system is used to test the imaging capability of graphene photodetectors.

**Keywords** Graphene · Photothermoelectric effect
Scanning photocurrent microscopy · Active infrared imaging · Resolution

## 2.1 Introduction

As the key components of infrared (IR) imaging systems, IR detectors have been improved greatly in the past century because of their important applications in the military, civilian and astronomy fields [1–3]. Various types of IR detectors have been developed based on narrow bandgap semiconductors like ternary alloys, group III-V and group IV semiconductors. The advanced technologies of material growth and fabrication processing contribute to the research of IR detectors [1, 2]. However, a key problem remains on how to achieve broadband absorption and fast response [4]. In the past decade, nano-materials with intriguing optoelectronic characteristics [5–10] have shown great potential for the application of future IR detectors to overcome these obstacles.

Graphene, as the first reported two-dimensional (2D) material, has drawn much attention in optoelectronic devices due to its unique properties such as broadband photon absorption [7, 8], ultrafast carrier dynamics [9] and high mobility [10]. Devices based on different physical mechanisms, including the photovoltaic (PV) effect [11–14], photothermoelectric (PTE) effect [15–17], photogating effect [18, 19] and bolometric effect [20], have been investigated to develop high-performance graphene photodetectors. High photoconductive gain and fast response have been achieved in graphene-based hybrid phototransistors [18, 19, 21], waveguide integration [22, 23], van der Waals heterostructure [24–26] and other device



N. Guo, *Infrared Photodetectors Based on Low-Dimensional Materials*,
Springer Theses, https://doi.org/10.1007/978-981-13-2838-1_2

configurations. Most of the researches were centered on device design, test and mechanism analysis, but there has been a growing demand for an implementation of high-resolution imaging from graphene photodetectors, especially in the IR spectral region.

In this chapter, an active IR imaging was carried out by a multilayer graphene photodetector at room temperature. High-resolution optical images with close-to-theoretical-limit Michelson contrast were obtained in the visible and short-wavelength IR spectral range. With Raman measurement and numerical simulation, the study has revealed that the photovoltaic and photothermoelectric effects are the main factors for the generation of photocurrents, contributing to high-quality images.

## 2.2 Results and Discussions

### *2.2.1 Device Structure and Photoresponse Measurement*

The graphene used in this work was mechanically exfoliated from a graphite flake onto a 300 nm $SiO_2$/Si substrate. The source/drain (S/D) electrodes (15 nm Cr and 50 nm Au) were prepared by electron-beam lithography, metallization, and lift-off process. To investigate the photoresponse mechanism, the device was characterized by scanning photocurrent microscopy (SPM) (see method). It should be noted that all the photocurrent (PC) maps were taken at zero source-drain bias. Figure 2.1a, b shows an illustration of the graphene photodetector and correlated spatially resolved PC map under 1550 nm light illumination. The channel material of multilayer graphene has a length of ~13 $\mu$m. Four red-and-blue PC ribbons appear across the graphene channel by scanning the focused laser spot over the device. Red and blue ribbons indicate the opposite polarity of PC. The PC, as a function of distance, along the white arrow marked in Fig. 2.1b is shown in Fig. 2.1c. There are two PC maxima near the S/D contacts, while there are two PC extrema in the middle of the conductive channel. These four PC values are of the same order of magnitude. And, the PCs increase linearly with optical power (see Fig. 2.1d). Many research groups have reported that the PCs occurring in close proximity to the contacts (corresponding to points A and D) are attributed to the separation of photogenerated electron-hole pairs by the PV or PTE effect [27, 28]. For the device with homogeneous graphene channel, PCs at a certain distance away from contacts disappear under zero applied bias, while our experiment showed otherwise. Therefore, the following discussion will focus on the positive and negative PC pair located at the channel center (corresponding to points B and C). In addition, Fig. 2.2 shows the transfer curve of graphene transistor. The fact that the $I_{ds}$ monotonously decrease with the increase of gate voltage $V_{gs}$ demonstrates that the graphene has a p-type characteristic.

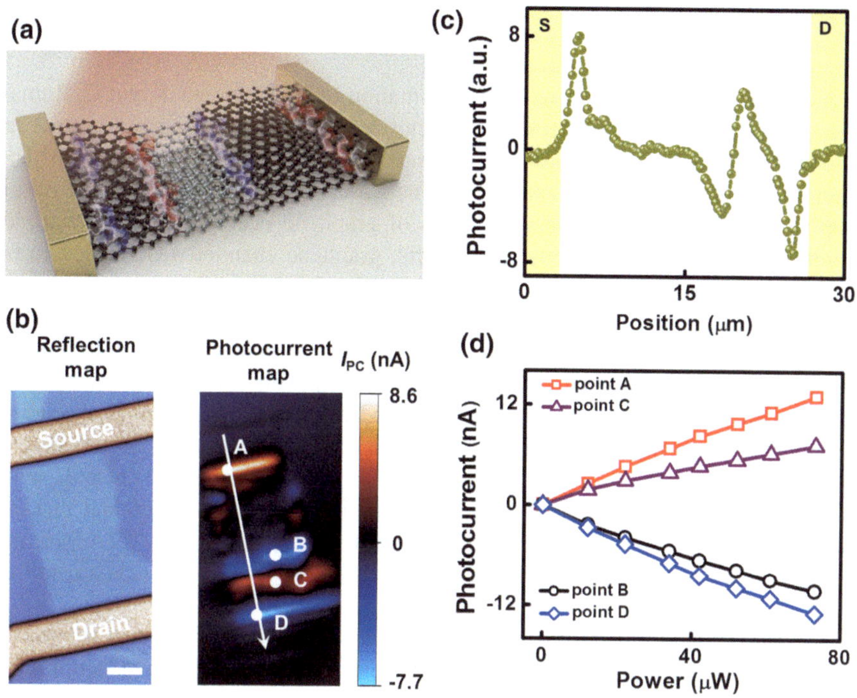


**Fig. 2.1** **a** A schematic illustration of the graphene photodetector. **b** Optical reflection map taken under 633 nm laser illumination and scanning PC map taken under 1550 nm laser illumination with light power of 22.5 μW. Scale bar, 3 μm. **c** PC line trace along the white arrow in (**b**). The shaded area indicates the contour of source/drain electrode. **d** PCs versus incident light power taken at the points A, B, C, and D in (**b**)

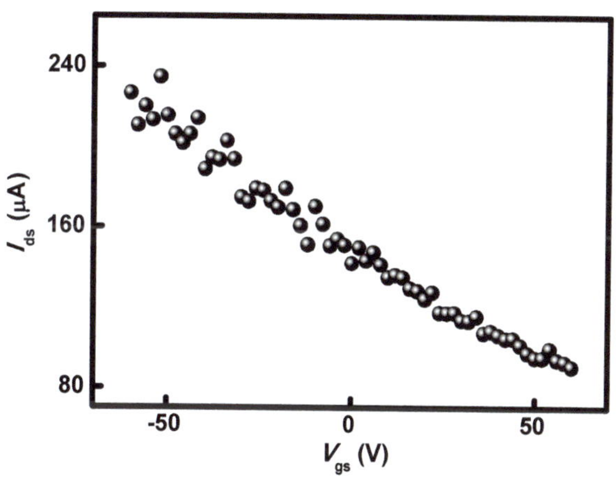


**Fig. 2.2** $I_{ds}$-$V_{gs}$ characteristic of the graphene transistor acquired at $V_{ds} = 0.1$ V

### 2.2.2 Physical Mechanism

To study the underlying reasons for the formation of PCs at points B and C, Raman microspectroscopy was employed to characterize the graphene property. Figure 2.3a, b depicts the Raman G(2D) band intensity maps and G(2D) band frequency maps. The Raman intensity maps have a contrast profile similar to that of the optical reflection map in Fig. 2.1b. Except a small portion of graphene near the source electrode, which is a little thicker than the other parts, graphene channel was verified to be uniform by the homogeneous G/2D band intensity Raman maps, especially for the PC region marked by a rectangle. Distinct blue shifts of G and 2D bands can be observed in Raman frequency maps, indicating a heavier p-doping in the marked graphene area compared with its neighbors [29–31]. This shift could be attributed to an unintentional doping by species absorbed from the surroundings or residual polymers used during the device fabrication [31]. Figure 2.3c displays the Raman spectra taken from the points as labeled by circles in Fig. 2.3a, which presents same intensity ratio $I_G/I_{2D}$ in p and p+ regions. This further testifies to the fact that the marked graphene region has uniform multilayer structure [32–34]. Therefore, the generation of positive and negative PC pair located at points B and C is a result of the p-p+-p junction.

As mentioned above, the PC generation at the graphene-metal contact is induced by PV or PTE effect. Due to the identical PC polarity and comparable PC amplitude, it is difficult to distinguish which effect could be the dominant response mechanism [15, 17, 20]. Graphene p-n junction also faces the similar problem. For unipolar junctions such as n-n+ and p-p+, these two effects will lead to different PC polarities

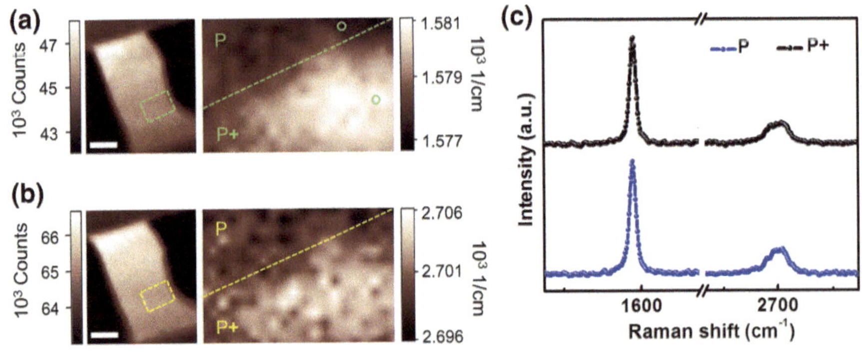


**Fig. 2.3** **a** Raman map of the G band intensity (left column) and frequency (right column, taken from dotted box). Scale bar, 3 μm. **b** Raman map of the 2D band intensity (left column) and frequency (right column, taken from dotted box). Scale bar, 3 μm. **c** Raman spectra corresponding to the points as labeled by circle in (**a**)

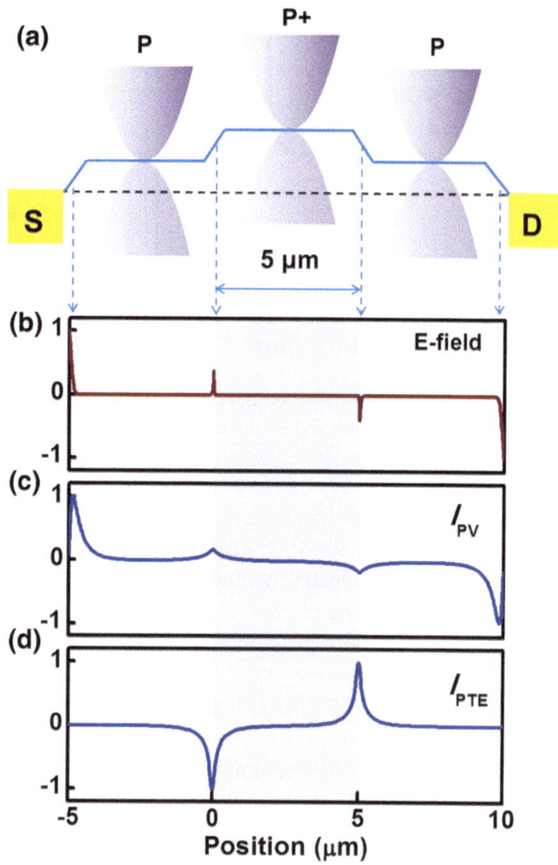


**Fig. 2.4** **a** Energy band structure of the graphene device. **b–d** Simulated distribution of the electric field and PCs. The values are normalized and the shaded area indicates the enlarged p+ region

[15, 20]. Different from experimental methods previously reported, our study explored the photoresponse in p-p+-p junction by numerical simulation. For the sake of simplicity, only PV effect was considered as the factor responsible for PCs at the graphene-metal junction.

An energy band structure of the graphene device based on the results of Raman measurements is given in Fig. 2.4a. The model used here is based on drift-diffusion approximation, where the Poisson's equation and continuity equations for electrons and holes were solved self-consistently by the finite element method [35, 36]. To simplify the situation, the multilayer graphene was regarded as a narrow-bandgap material (0.2 eV) and its dimension was enlarged for convergence during the simulation [37]. The Schottky barrier at the graphene-metal junction was set to 0.2 eV. The p region doping concentration of $3.08 \times 10^{13}$ cm$^{-2}$ can be acquired from the $I_{ds}$-$V_{gs}$ characteristic in Fig. 2.2 at $V_{gs} = 0$ V. Because accurately defining the doping concentration of p+ region is quite difficult, one approximate approach was adopted using the hysteresis behavior of transfer curve in graphene transistor to achieve a slight p+ doping concentration of $3.8 \times 10^{13}$ cm$^{-2}$. This approximation will be explained

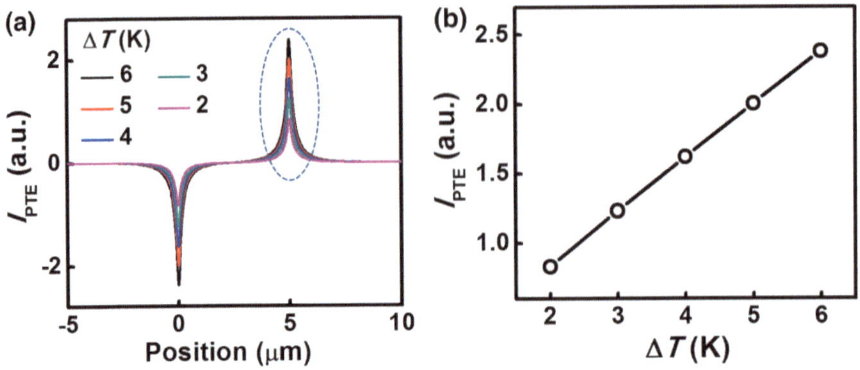


**Fig. 2.5** **a** Simulated distribution of $I_{PTE}$ for the temperature difference $\Delta T$ ranging from 2 to 6 K between the junction area and its surroundings. **b** Simulated $I_{PTE}$, taken from the labeled peaks in (**a**), versus $\Delta T$

below in detail. For the PTE effect, a temperature difference induced by laser spot led to a net thermoelectric voltage across the graphene channel. After a fast initial relaxation process, the photoexcited carriers would form a hot Fermion distribution. Driven by temperature gradient, hot carriers produced a thermal current that forms the thermoelectric current [15, 17, 28]. The simulation used a thermodynamic model to account for the PTE effect. The temperature gradient as a driving term was added in the drift-diffusion model. Figure 2.4b, c gives the simulation results of electric field distribution along the conductive channel and corresponding PC induced by PV effect ($I_{PV}$). It is obvious that the directions of $I_{PV}$ at the p-p+-p junction are opposite to that of PC obtained from experiment ($I_{Exp}$), which proves that the PV effect is not the reason for the generation of $I_{Exp}$ at the p-p+-p junction. Figure 2.4d shows the simulated distribution of PC induced by PTE effect ($I_{PTE}$). It can be seen that the direction of $I_{PTE}$ is consistent with that of $I_{Exp}$. Moreover, Fig. 2.5 shows the simulated $I_{PTE}$ with the temperature difference ranging from 2 to 6 K, where a good linear dependence of $\Delta T$ was obtained. Therefore, the PTE mechanism leads to the PC generation at p-p+-p junction.

### 2.2.3 $V_{gs}$ *Dependence of Photoresponse*

In order to study how PC changes with $V_{gs}$ modulation, SPM was used to investigate the photoresponse with different $V_{gs}$. Figure 2.6 presents the PC maps of another device in our work, which shows that PC map with 565 nm illumination can provide more details than 1550 nm illumination does because light with short wavelengths would result in a smaller diffraction-limited light spot. Therefore, in Fig. 2.7a, 565 nm light is used to get high-resolution PC maps at $V_{gs}$ of −60, 0 and 60 V. The PC distribution and morphology are basically same as that shown in Fig. 2.1b.

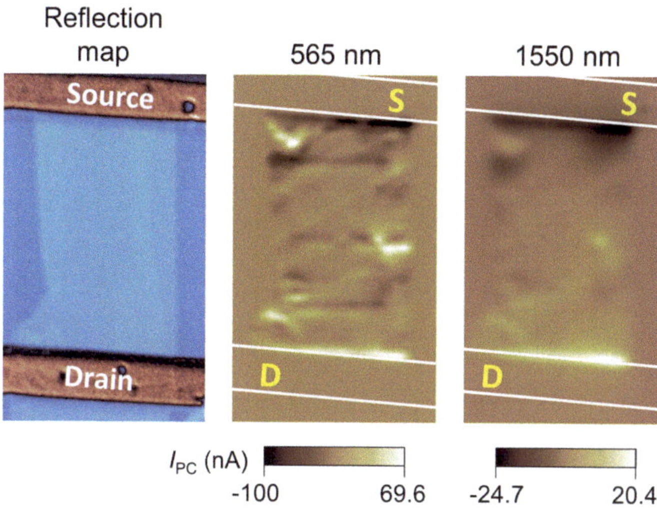


**Fig. 2.6** Another device's optical reflection map and scanning PC maps taken under 565 and 1550 nm light illumination

For $I_{PV}$, the current amplitude decreases with the increase of $V_{gs}$ (see Fig. 2.7b), which can be explained by the fact that when graphene's Fermi level approaches Dirac point, the strength of built-in electric field at the graphene-metal junction is gradually weakened (see the inset of Fig. 2.7b). It should be noted that there is a maximum value for $I_{PTE}$ as $V_{gs}$ increases at points B and C (see Fig. 2.7c). Usually, the direction and magnitude of $I_{PTE}$ depend on the thermoelectric power (Seebeck coefficient) in each part of graphene channel and can be expressed as [17]

$$I_{PTE} = \frac{S_2 - S_1}{R} \Delta T \tag{2.1}$$

where $S$ is the Seebeck coefficient in the p and p+ regions, and $\Delta T$ is the temperature difference induced by laser illumination. From the Mott relation [38, 39], the Seebeck coefficient can be written as

$$S = \frac{\pi^2 k_b^2 T}{3e} \frac{1}{R} \frac{dR}{dV_{gs}} \frac{dV_{gs}}{dE}\bigg|_{E=E_f} \tag{2.2}$$

where $k_b$ is the Boltzmann constant, $e$ is electron charge, $T$ is the sample temperature, $R$ is the resistance, $dR/dV_{gs}$ can be derived from the transport characteristic measurements, and $E_f$ is the Fermi energy. $E_f(n)$ dependence of the multilayer graphene in this work was approximated by that of bilayer graphene, which is formulated as [17, 25, 40]

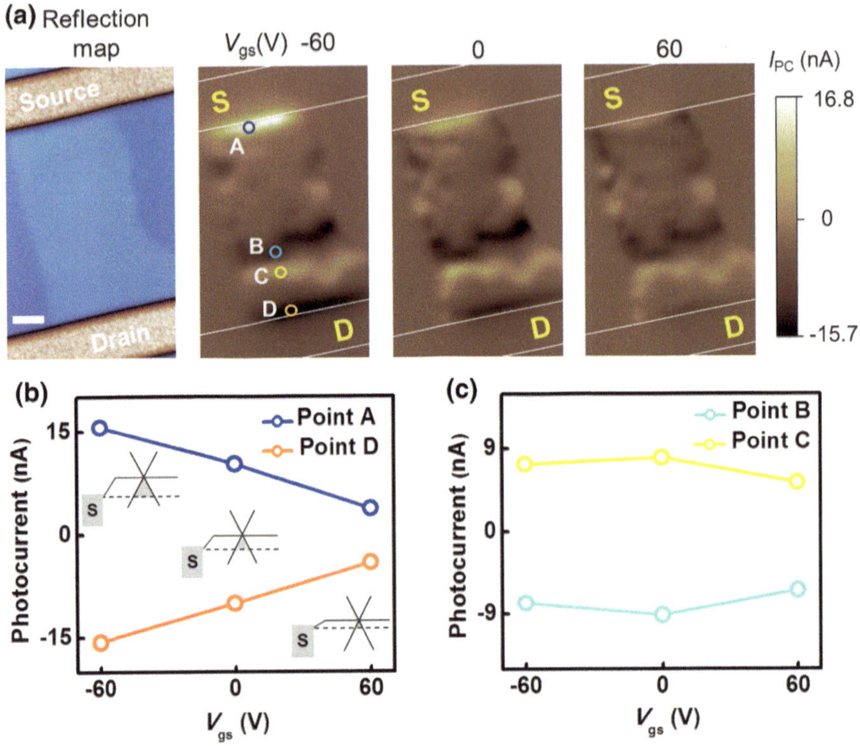


**Fig. 2.7** **a** Optical reflection map and scanning PC maps taken under $V_{gs}$ of −60, 0 and 60 V with 565 nm light illumination (18.2 μW). Scale bar, 2 μm. **b** The PCs, taken at points A and D, versus $V_{gs}$. The insets are the energy band alignment with $V_{gs}$ modulation. **c** The PCs, taken at points B and C, versus $V_{gs}$

$$E_f = \frac{1}{2}\sqrt{(2\hbar v_F)^2 \pi n + 2\gamma_1^2 - 2\gamma_1\sqrt{(2\hbar v_F)^2 \pi n + \gamma_1^2}} \tag{2.3}$$

where $\gamma_1$ is the interlayer coupling strength and set to be 0.4 eV [25, 41], $\hbar$ is Planck's constant, and the Fermi velocity $v_F = 10^6$ m/s [38].

To explain the phenomenon in Fig. 2.7c, Eqs. (2.1), (2.2) and (2.3) should be solved together, but extracting accurate values of $S_{\text{p}}$ and $S_{\text{p+}}$ for p and p+ regions in the same conductive channel is very difficult. Therefore, our study adopted an approximate approach to get p- and p+-type channel on the same device, that is, by changing the sweeping direction of gate bias from forward to backward a slight hysteresis in transfer curve is obtained. Many research groups have investigated the hysteresis behavior in graphene transistor, and their explanation is that charge injection into trap sites on a dielectric substrate and charge transfer from/to neighboring adsorbates can result in the hysteresis [42, 43]. Figure 2.8a shows the transfer curve of another device in our work. The black and red dotted lines reveal the p- and p+-type graphene

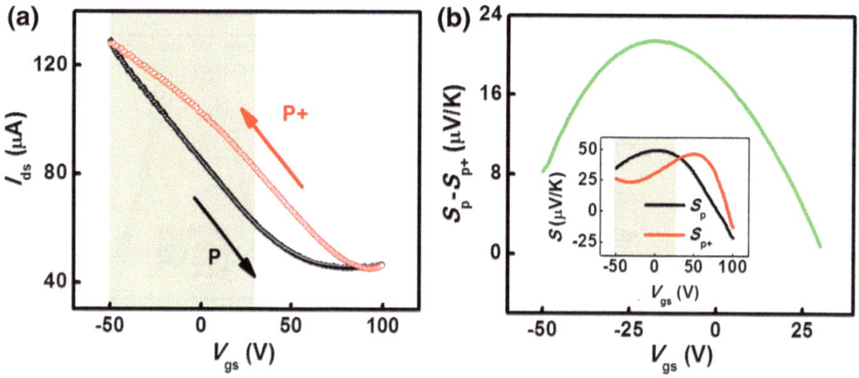


**Fig. 2.8** **a** $I_{ds}$-$V_{gs}$ characteristic of another graphene transistor acquired at $V_{ds} = 0.1$ V. The arrows label the $V_{gs}$ sweeping direction. **b** The calculated thermoelectric power based on the transfer property in (**a**), versus $V_{gs}$

characteristics, respectively. Here, the transport characteristic in the $V_{gs}$ range of −50 to 30 V, marked with shaded area, was chosen to analyze the $I_{PTE}$ change with $V_{gs}$ due to its similarity in line shape to the curve in Fig. 2.2. Figure 2.8b presents the calculated $S_p$ and $S_{p+}$ (the inset) and the difference ($S_p - S_{p+}$). Compared with $I_{PTE}$ in Fig. 2.7c, the difference in Seebeck coefficient shows a consistent variation tendency with $V_{gs}$ modulation. It should be noted that only the amplitude of $I_{PTE}$ changes with $V_{gs}$ and no polarity reversal is observed in Fig. 2.7a, which is different from the calculated results (unshaded area in the inset of Fig. 2.8b) in which a pronounced change in the $I_{PTE}$ sign is predicted, that is, the sign of ($S_p - S_{p+}$) at high $V_{gs}$. This may be accounted for by the fact that the reversal in the $I_{PTE}$ sign happened when $V_{gs}$ was approaching the Dirac point as shown in the inset of Fig. 2.8b. In the case of Fig. 2.7a, however, the graphene is more p-doped than that in Fig. 2.8a, and its $V_{gs}$ is not high enough to tune the Fermi level to the Dirac point.

### 2.2.4 *Multispectral Active Infrared Imaging*

A confocal laser scanning microscope setup was refitted to investigate the imaging ability of the graphene device, which was configured as a point-like detector to collect the reflected light of the illuminated objects (see Fig. 2.9a). The signal PCs can be extracted from the graphene device to form images through the current pre-amplifier and lock-in amplifier. To achieve the best performance, a × 100 objective lens 1 with a numerical aperture of 0.95 is used to obtain a close-to-diffraction-limited light spot on the objects. Then, the graphene detector is slightly moved in three dimensional directions near the focus of objective lens 2 (×20/NA-0.4) to achieve maximum signal currents. The PCs consist of $I_{PTE}$ and $I_{PV}$, which are produced by hybrid response mechanisms, i.e. the synergistic action of PTE and PV effects. Figure 2.9b

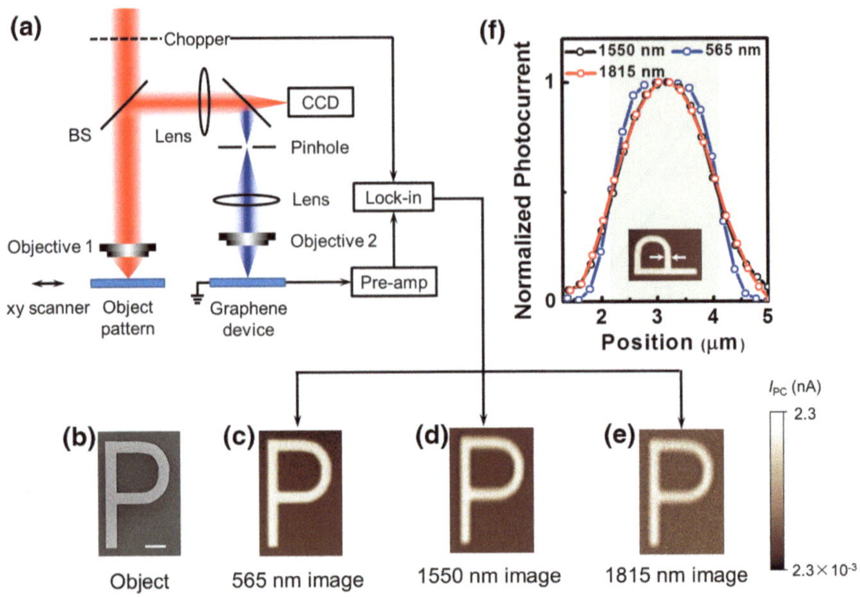


**Fig. 2.9** **a** Schematic of the active imaging setup. **b** Scanning electron microscopy image of metallic objects (P) with a line width of 2 μm patterned on a silicon substrate. Scale bar, 4 μm. **c–e** Images acquired from graphene photodetector at λ of 565, 1550 and 1815 nm. **f** The signal PCs taken from a pattern edge in images. The shaded area indicates the contour of pattern

gives a scanning electron microscopy image of metallic objects (P) with a line width of 2 μm patterned on a silicon substrate. The corresponding images are acquired from the graphene detector by applying the focused laser spot (565, 1550 and 1815 nm) scanning across the patterns (see Fig. 2.9c–e). The signal PCs from a pattern edge in images are extracted (see Fig. 2.9f). The resolution of the 565 nm image is slightly higher than that of the other two images (1550 and 1815 nm), which have a similar profile. The reason is that for the active infrared imaging system in our work, the key factor influencing the resolution is objective lens 1. When a shorter wavelength light passes through an objective, a smaller diffraction-limited light spot will be focused onto the objects, producing a relatively clearer image outline than a light with longer wavelength does. In order to obtain the image resolution, the model of point spread function based on the convolution of a step function and a Gaussian function [44] are used to fit the data in Fig. 2.9f. As shown in Fig. 2.10, the optical resolutions of 418, 657 and 877 nm are extracted from the fitting functions for the 565, 1550 and 1815 nm images, respectively. The results prove graphene photodetector's good performance in the broadband high-resolution imaging applications.

Furthermore, in order to characterize the image quality, the Michelson contrast $C$ is calculated with formula [44, 45]

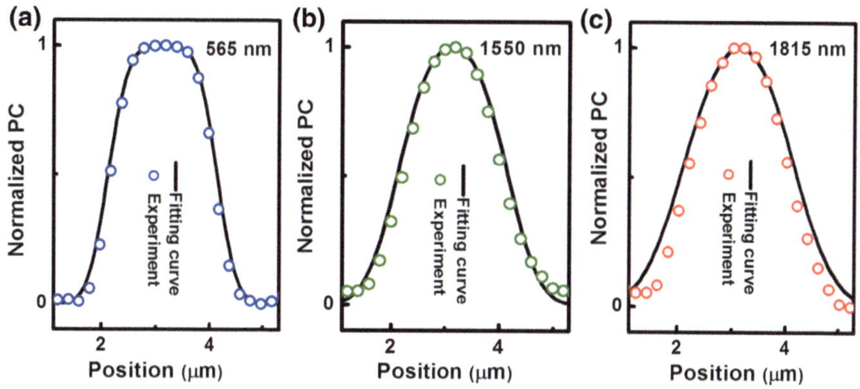


**Fig. 2.10** The model of point spread function based on a convolution of a step function and a Gaussian function used to fit the data in Fig. 2.9f. The optical resolutions of 418, 657 and 877 nm are extracted from the fitting functions for the 565 nm (**a**), 1550 nm (**b**) and 1815 nm (**c**) images, respectively

$$C = \frac{L_{max} - L_{min}}{L_{max} + L_{min}} \tag{2.4}$$

where $L_{max}$ ($L_{min}$) is the PC maximum (minimum) of the images. As a result, ultrahigh contrasts of 0.997, 0.994 and 0.996 are obtained for 565, 1550 and 1815 nm images, respectively. The graphene photodetector shows good performance for imaging with close-to-theoretical-limit contrast in the visible and short-wavelength infrared spectral range.

## 2.3 Conclusions

An active infrared imaging by multilayer graphene photodetectors with hybrid response mechanisms is demonstrated at room temperature. The high optical resolutions of 418, 657 and 877 nm and close-to-theoretical-limit Michelson contrasts of 0.997, 0.994, and 0.996 have been acquired in the visible (565 nm) and short-wavelength infrared (1550 and 1815 nm) spectral range, respectively. Detailed studies based on the Raman measurements and numerical simulation of response process reveal that hybrid photocurrents in multilayer graphene detector can be attributed to the synergistic action of photovoltaic and photothermoelectric effects. These results will help promote the development of high performance graphene-based IR multispectral detectors.

## 2.4 Detailed Methods

The optoelectronic properties of our devices were characterized by scanning photocurrent microscopy (SPM). SPM consists of optical module, scanning module and detection module. The optical module is responsible for realizing confocal system. Scanning module is used to move the sample in X-Y plane. Finally, the photocurrent is collected by detection module. It is noteworthy that, in this SPM measurements, a femtosecond OPO laser (Spectra-Physics Inspire ultrafast OPO system with tunable wavelength of 345–2500 nm) was used as the excitation source. The graphene device is mounted onto a piezo-driven sample stage, and then illuminated by a focused laser spot using a ×100/NA-0.95 objective. The laser beam is modulated at the frequency ~893 Hz with a mechanical chopper used as reference to a lock-in amplifier from which the photocurrent signal can be extracted. Simultaneously, the reflection map is obtained through collecting the reflected light by a silicon avalanche photodiode (APD). For active imaging measurements, the APD was replaced by the graphene device as a point-like photodetector to collect the reflected light of illuminated objects. The objects made of metal were patterned on the silicon substrate which is mounted onto the piezo-driven sample stage. The scan speed is 25.2 ms/pixel and there are 5 pixels per micron. A ×100/NA-0.95 objective was used to focus the laser spot on the patterns. At the same time, a ×20/NA-0.4 objective is positioned in front of the graphene photodetector for better light collection. Raman spectra were measured by Nanofinder 30 (TII Tokyo Instruments, Inc.). All spectra are obtained by a solid state laser ($\lambda = 532$ nm) with 0.5 mW excitation power, and calibrated by the Raman shift of the single crystal silicon at 520.4 $cm^{-1}$. The incident laser is focused on the sample with a ×100/NA-0.7 objective.

## References

1. W. Lei, J. Antoszewski, L. Faraone, Appl. Phys. Rev. **2**, 041303 (2015)
2. A. Rogalski, Infrared Phys. Technol. **54**, 136 (2011)
3. A. Rogalski, Prog. Quantum Electron. **27**, 59 (2003)
4. N. Liu, H. Tian, G. Schwartz, J.B.-H. Tok, T.-L. Ren, Z. Bao, Nano Lett. **14**, 3702 (2014)
5. N. Youngblood, C. Chen, S.J. Koester, M. Li, Nat. Photonics **9**, 247 (2015)
6. H. Yuan, X. Liu, F. Afshinmanesh, W. Li, G. Xu, J. Sun, B. Lian, A.G. Curto, G. Ye, Y. Hikita, Z. Shen, S.-C. Zhang, X. Chen, M. Brongersma, H.Y. Hwang, Y. Cui, Nat. Nanotechnol. **10**, 707 (2015)
7. J.M. Dawlaty, S. Shivaraman, J. Strait, P. George, M. Chandrashekhar, F. Rana, M.G. Spencer, D. Veksler, Y. Chen, Appl. Phys. Lett. **93**, 131905 (2008)
8. M. Liu, X. Yin, E. Ulin-Avila, B. Geng, T. Zentgraf, L. Ju, F. Wang, X. Zhang, Nature **474**, 64 (2011)
9. J.M. Dawlaty, S. Shivaraman, M. Chandrashekhar, F. Rana, M.G. Spencer, Appl. Phys. Lett. **92**, 042116 (2008)
10. C.R. Dean, A.F. Young, I. Meric, C. Lee, L. Wang, S. Sorgenfrei, K. Watanabe, T. Taniguchi, P. Kim, K.L. Shepard, J. Hone, Nat. Nanotechnol. **5**, 722 (2010)
11. T. Mueller, F. Xia, P. Avouris, Nat. Photonics **4**, 297 (2010)
12. T. Mueller, F. Xia, M. Freitag, J. Tsang, P. Avouris, Phys. Rev. B **79**, 245430 (2009)

13. M.C. Lemme, F.H.L. Koppens, A.L. Falk, M.S. Rudner, H. Park, L.S. Levitov, C.M. Marcus, Nano Lett. **11**, 4134 (2011)
14. E.C. Peters, E.J.H. Lee, M. Burghard, K. Kern, Appl. Phys. Lett. **97**, 193102 (2010)
15. N.M. Gabor, J.C.W. Song, Q. Ma, N.L. Nair, T. Taychatanapat, K. Watanabe, T. Taniguchi, L.S. Levitov, P. Jarillo-Herrero, Science **334**, 648 (2011)
16. J.C.W. Song, M.S. Rudner, C.M. Marcus, L.S. Levitov, Nano Lett. **11**, 4688 (2011)
17. X. Xu, N.M. Gabor, J.S. Alden, A.M. van der Zande, P.L. McEuen, Nano Lett. **10**, 562 (2010)
18. G. Konstantatos, M. Badioli, L. Gaudreau, J. Osmond, M. Bernechea, F.P.G. de Arquer, F. Gatti, F.H.L. Koppens, Nat. Nanotechnol. **7**, 363 (2012)
19. C.-H. Liu, Y.-C. Chang, T.B. Norris, Z. Zhong, Nat. Nanotechnol. **9**, 273 (2014)
20. M. Freitag, T. Low, F. Xia, P. Avouris, Nat. Photonics **7**, 53 (2013)
21. Y. Liu, F. Wang, X. Wang, X. Wang, E. Flahaut, X. Liu, Y. Li, X. Wang, Y. Xu, Y. Shi, R. Zhang, Nat. Commun. **6**, 8589 (2015)
22. A. Pospischil, M. Humer, M.M. Furchi, D. Bachmann, R. Guider, T. Fromherz, T. Mueller, Nat. Photonics **7**, 892 (2013)
23. X. Gan, R.-J. Shiue, Y. Gao, I. Meric, T.F. Heinz, K. Shepard, J. Hone, S. Assefa, D. Englund, Nat. Photonics **7**, 883 (2013)
24. W.J. Yu, Y. Liu, H. Zhou, A. Yin, Z. Li, Y. Huang, X. Duan, Nat. Nanotechnol. **8**, 952 (2013)
25. F. Withers, T.H. Bointon, M.F. Craciun, S. Russo, ACS Nano **7**, 5052 (2013)
26. M. Massicotte, P. Schmidt, F. Vialla, K.G. Schädler, A. Reserbat-Plantey, K. Watanabe, T. Taniguchi, K.J. Tielrooij, F.H.L. Koppens, Nat. Nanotechnol. **11**, 42 (2016)
27. J. Park, Y.H. Ahn, C. Ruiz-Vargas, Nano Lett. **9**, 1742 (2009)
28. T.J. Echtermeyer, P.S. Nene, M. Trushin, R.V. Gorbachev, A.L. Eiden, S. Milana, Z. Sun, J. Schliemann, E. Lidorikis, K.S. Novoselov, A.C. Ferrari, Nano Lett. **14**, 3733 (2014)
29. J. Yan, Y. Zhang, P. Kim, A. Pinczuk, Phys. Rev. Lett. **98**, 166802 (2007)
30. W.H. Wang, H.Y. Nan, Q. Liu, Z. Liang, Z.H. Yu, F.Y. Liu, W.D. Hu, W. Zhang, X.R. Wang, Z.H. Ni, Appl. Phys. Lett. **106**, 021121 (2015)
31. H. Liu, Y. Liu, D. Zhu, J. Mater. Chem. **21**, 3335 (2011)
32. Z. Ni, Y. Wang, T. Yu, Z. Shen, Nano Res. **1**, 273 (2008)
33. A.C. Ferrari, J.C. Meyer, V. Scardaci, C. Casiraghi, M. Lazzeri, F. Mauri, S. Piscanec, D. Jiang, K.S. Novoselov, S. Roth, A.K. Geim, Phys. Rev. Lett. **97**, 187401 (2006)
34. Y. Gong, X. Zhang, G. Liu, L. Wu, X. Geng, M. Long, X. Cao, Y. Guo, W. Li, J. Xu, M. Sun, L. Lu, L. Liu, Adv. Funct. Mater. **22**, 3153 (2012)
35. X. Li, W. Chen, S. Zhang, Z. Wu, P. Wang, Z. Xu, H. Chen, W. Yin, H. Zhong, S. Lin, Nano Energy **16**, 310 (2015)
36. W. Chen, A. Rinzler, J. Guo, J. Appl. Phys. **113**, 094507 (2013)
37. E. Sano, T. Otsuji, Appl. Phys. Express **2**, 061601 (2009)
38. Y.M. Zuev, W. Chang, P. Kim, Phys. Rev. Lett. **102**, 096807 (2009)
39. M. Cutler, N.F. Mott, Phys. Rev. **181**, 1336 (1969)
40. A.H. Castro Neto, F. Guinea, N.M.R. Peres, K.S. Novoselov, A.K. Geim, Rev. Mod. Phys. **81**, 109 (2009)
41. E. McCann, D.S.L. Abergel, V.I. Fal'ko, Solid State Commun. **143**, 110 (2007)
42. H. Wang, Y. Wu, C. Cong, J. Shang, T. Yu, ACS Nano **4**, 7221 (2010)
43. G. Kalon, Y. Jun Shin, V. Giang Truong, A. Kalitsov, H. Yang, Appl. Phys. Lett. **99**, 083109 (2011)
44. M. Engel, M. Steiner, P. Avouris, Nano Lett. **14**, 6414 (2014)
45. A.A. Michelson, *Studies in Optics* (University of Chicago Press, Chicago, IL, 1927)

# Chapter 3
# Highly-Efficient InAs Nanowire Phototransistors

**Abstract** This chapter describes an anomalous, high-gain and broadband photoresponse in InAs nanowire phototransistors. Photogating layer formed at the surface of InAs nanowires traps photocarriers, leading to a local electric field to modulate the conductance of nanowires. The strength and polarity of photogate will be studied by applying different back-gate voltages.

**Keywords** InAs nanowire · Phototransistor · Photogating effect

## 3.1 Introduction

Semiconductor nanowires (NWs) have attracted wide attention due to their potential applications in nano-electronics and nano-photonics in recent decades [1–22]. Many kinds of NWs have been grown and fabricated into devices to explore their optoelectronic characteristics. Surface states of NWs are expected to play an important role in photosensitivity owing to their large surface-to-volume ratio [21, 23]. What's more, a faster transport speed may be realized by shorter conductive channels [23]. Of all semiconductor NWs, InAs NW has proved to be an extraordinary material. Due to its narrow bandgap of ~0.354 eV [24] and high carrier mobility at room temperature [12], InAs NW is highly promising for applications in broadband spectrum detection, from visible to infrared regions. However, NW-based photodetectors suffer from a common problem: short lifetime of minority carrier, which lowers their responsivity. A significant number of minority carriers are recombined before being collected by the electrodes. Visible-blind UV photodetectors made by ZnO NWs is known to have an internal photoconductive gain of ~$10^8$ reported by Soci et al. [21], whose research shows that the hole lifetime can be prolonged by oxygen-related hole-trap states at NW surface, resulting in an ultrahigh photoconductive gain. Despite its excellent performance, device works well only in oxygen-rich environment and its effective spectral range is restricted to UV light. For most NW-based photodetectors reported, photocurrents mainly come from the junction area such as NW-metal junction [16, 25–29], p-n junction [30–32] and heterojunction [33–36]. The performance



N. Guo, *Infrared Photodetectors Based on Low-Dimensional Materials*,
Springer Theses, https://doi.org/10.1007/978-981-13-2838-1_3

of the device is very difficult to improve because of its small effective junction region and weak optical absorption, and the complex process of synthesis and device fabrication still pose great challenges for its wide application. Thus, in the development of high-sensitive and broadband spectrum NW detectors in the future, minority carrier transport mechanism and junction-type device structure may become two main obstacles to overcome.

In this study, core/shell-like n-type InAs NWs were grown by chemical vapor deposition (CVD). By controlling NW growth, a self-assembled "photogating layer" (PGL) was formed near the NW surface. The main function of the PGL was trapping electrons photoexcited from the core, which in return produced a built-in electric field to modulate the core conductance, i.e. photogating effect [37, 38]. In our design, electrons were majority carriers, which were responsible for the generation of photocurrents. Therefore, the whole channel of NW as effective photosensitive region responded to light signal. High photoconductive gain of ~$-10^5$ and fast response time of 12 ms were achieved at room temperature. Moreover, our device showed good performance in both air and vacuum environments compared with other NW-based detectors, in which the response largely depends on gas molecules. This mechanism of majority carrier-dominated photodetection provides a novel approach to developing NW-based photodetectors for high-sensitive, broadband spectrum detection applications.

## 3.2 Results and Discussions

### *3.2.1 Nanowire Growth and Device Fabrication*

A solid-source CVD method [39, 40] was used to prepare the InAs NWs. InAs powders (1.2 g; 99.9999% purity) were heated in the upstream source zone of a two-zone tube furnace and the evaporated precursors were transported by a carrier gas of $H_2$ (99.999% purity; 200 sccm) to the downstream substrate zone. For this two-step growth method, a 0.5 nm thick Ni catalyst film pre-deposited on $SiO_2$/Si substrates was employed and annealed at 800 °C for 10 min. For the whole process, the pressure was maintained at ~3.0 Torr. The substrate temperature was first lowered to the nucleation temperature of 560 °C in the downstream zone. The nucleation of NWs began when the source temperature reached the required temperature of 690 °C in the upstream zone. After 2 min, the downstream temperature was further lowered to the second step growth temperature of 470 °C and kept for 20 min. Finally, the substrates were naturally cooled to room temperature.

The scanning electron microscope (SEM) image of InAs NWs is shown in Fig. 3.1a. Figure 3.1b, c provide transmission electron microscopy (TEM) images of a typical NW. We can see from the low-magnification TEM image in Fig. 3.1b that the NW has an uneven surface. Also, from the high-resolution TEM image in Fig. 3.1c it can be seen that the NW has a well-crystallized core with a defected shell

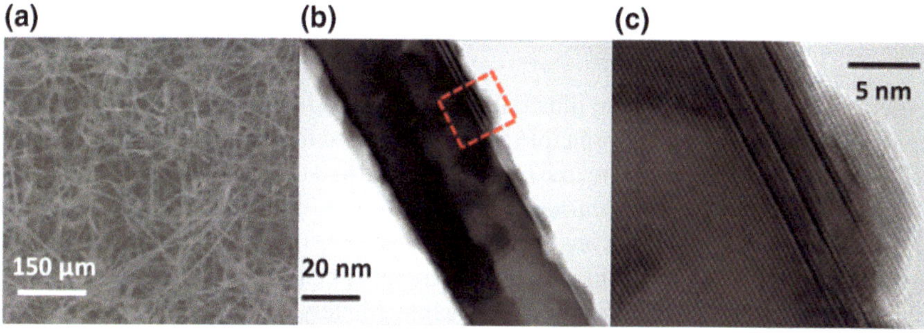


**Fig. 3.1** **a** SEM image of InAs NWs. **b** Bright-field TEM image of the InAs NW. **c** The high-resolution TEM image is taken from highlighted position in (**b**)

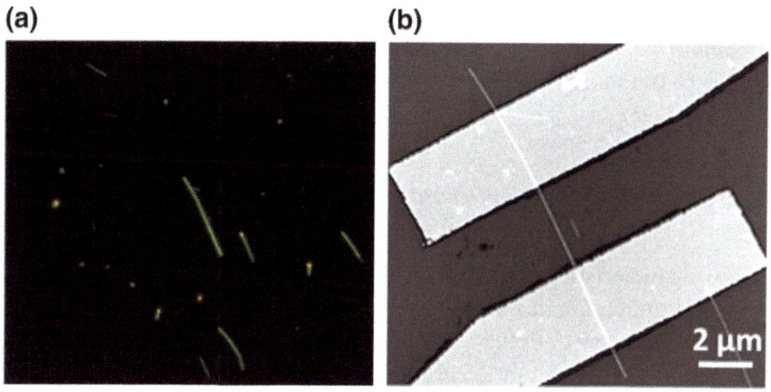


**Fig. 3.2** **a** Dark-field microscope image of InAs NW on $p^+$-Si/$SiO_2$ (300 nm) substrate. **b** SEM image of the as-fabricated InAs NW transistor

near the surface, i.e. the PGL. There are two possible explanations for the formation of defected shell in our case. The first one is that the source temperature (upstream zone) is relatively high (690 °C), which provides great volumes of source vapors to create catalytic supersaturation for NW growth, and the defected shell is induced by the non-uniform reaction of the excess vapor on NW surface. The second one is that the high process pressure ~3.0 Torr, which leads to shorter mean-free path of precursor vapor, which is conducive to the formation of defected shell.

For device fabrication, first, InAs NWs were transferred mechanically to pre-cleaned $p^+$-Si/$SiO_2$ (300 nm) substrates. A dark-field microscope image of InAs NW on $p^+$-Si/$SiO_2$ (300 nm) substrate is given in Fig. 3.2a. Then, the source/drain (S/D) patterns were defined by using electron-beam lithography. The NWs were dipped into a 2% HF solution for ~10 s before metallization in order to remove the native oxide to ensure the formation of Ohmic contact between the metal electrode (15 nm Cr/60 nm Au) and NW. Figure 3.2b presents a SEM image of the InAs NW transistor. As channel material, the InAs NW has a length of ~3.5 μm and a diameter of ~40 nm.

### 3.2.2 Physical Mechanism

Figure 3.3a gives a schematic illustration of the core/shell-like InAs NW transistor configuration. The working principle can be illustrated through the schematic shown in Fig. 3.3b, c. In NW core, the free electrons flow driven by electric field before light illumination to form the dark current, $I_{ds} > 0$ (see Fig. 3.3b). Under light illumination, the photogenerated electrons are excited from the core into the PGL (see Fig. 3.3b), where there are many randomly distributed trapping centers, leaving unpaired holes to recombine with free electrons in the core immediately (see Fig. 3.3c). At the same time, a built-in electric field was induced by the electrons trapped in the PGL to further deplete free electrons in the core through capacitive coupling (see Fig. 3.3c). The synergistic effect of these two factors leads to a great decrease in photocurrent, $I_{ds} \approx 0$. As a matter of fact, the core/shell-like NW has the identical composition of InAs, but the shell, i.e. the PGL, contains many randomly distributed lattice defects which serve as trapping centers to capture photoexcited electrons.

### 3.2.3 Photoresponse Performance

The $I_{ds}$-$V_{ds}$ characteristics of the InAs NW transistor measured in the dark and under a green laser illumination in air are given in Fig. 3.4a. The $I_{ds}$-$V_{ds}$ curve measured in the dark shows the conventional long-channel transistor characteristics, i.e. a linear regime at low $V_{ds}$ and a saturation regime at high $V_{ds}$ [41]. The $I_{Dark}$ increases with $V_{gs}$, indicating an n-type characteristic for NW. The $I_{Light}$ decreases significantly to a very low level upon light exposure, in which the net photocurrent, defined as $I_{PC} = |I_{Light}| - |I_{Dark}|$, as high as $-4.5$ μA is obtained at $V_{gs} = 40$ V and $V_{ds} = 1$ V (see Fig. 3.4b). For the conventional photodetector with positive photoresponse, the photoconductive gain ($G$), defined as the number of charges collected by electrodes due to the excitation by one photon, can be expressed as [20] $G = (I_{PC}/P)(h\nu/e)$,

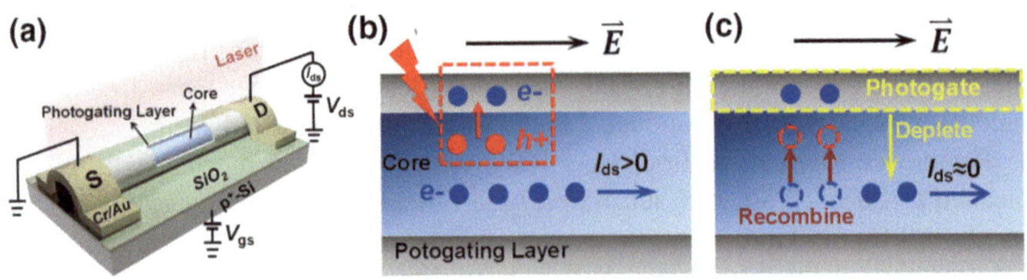


**Fig. 3.3** **a** A schematic illustration of the InAs NW transistor. **b** The process of photogenerated electron-hole pairs. Electrons ($e-$) are excited into the PGL and remain trapped under light illumination. Holes ($h+$) are left in the NW core. **c** The process of majority carrier-dominated photodetection. Holes recombine with free electrons in the core. Electrons trapped in PGL lead to a photogating effect on the core

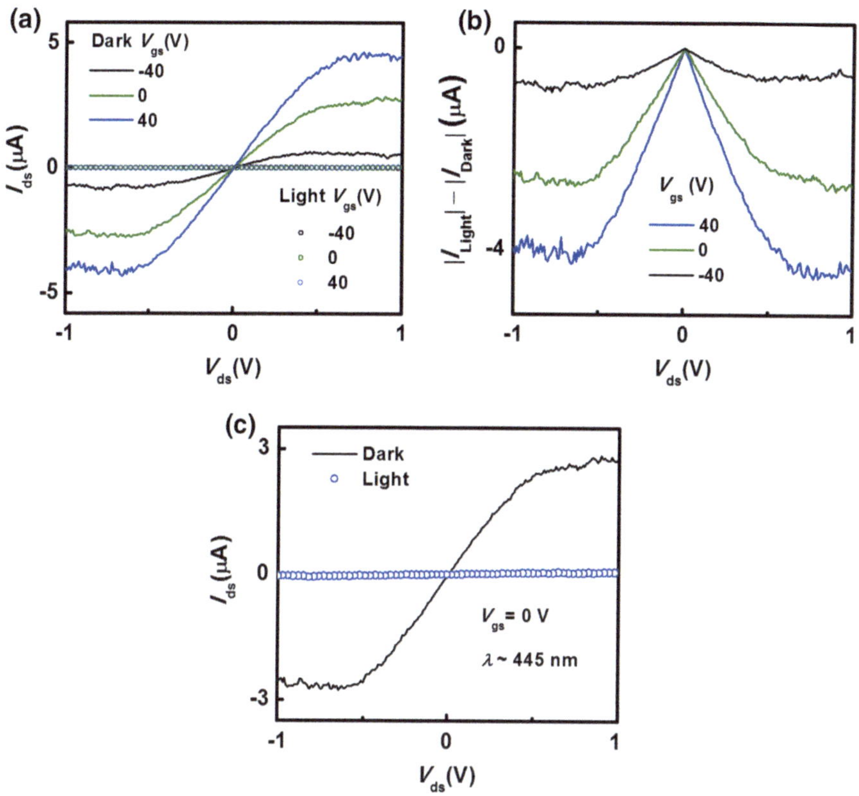


**Fig. 3.4** **a** $I_{ds}$-$V_{ds}$ curves of the InAs NW transistor in the dark and under illumination in air recorded for different gate voltages. The wavelength is 532 nm and light intensity is 8 mW mm$^{-2}$. **b** The net photocurrent, defined as $|I_{Light}| - |I_{Dark}|$, versus $V_{ds}$. **c** $I_{ds}$-$V_{ds}$ curve of the InAs NW transistor under 445 nm illumination in air at $V_{gs} = 0$ V. The light intensity is 8 mW mm$^{-2}$

where $h\nu$ is the energy of the incident photon, $e$ is the electron charge, and $P$ is the light power absorbed by the NW. However, for the anomalous photoresponse based on majority carrier transport in our work, the definition of negative photoconductive gain is how many carriers "cannot be collected" by electrodes for per photon incident. This definition is used for convenience of comparison with other NW detectors with positive photoresponse. Assuming that light incident on the NW is absorbed completely, an ultrahigh negative gain of ~$-10^5$ is obtained at $V_{gs} = 40$ V and $V_{ds} =$ 1 V. As anticipated in our designed experiments, this anomalous phenomenon stems from majority carrier-dominated photodetection mechanism. The $I_{ds}$-$V_{ds}$ characteristic measured under a blue laser illumination is shown in Fig. 3.4c. The large decrease of photocurrent is comparable to that in Fig. 3.4a.

Temporal response of InAs NW transistor under different $V_{gs}$ was investigated to understand how the strength of photogate changed with $V_{gs}$ modulation. The photoconductance modulation is realized by switching on and off the incident light

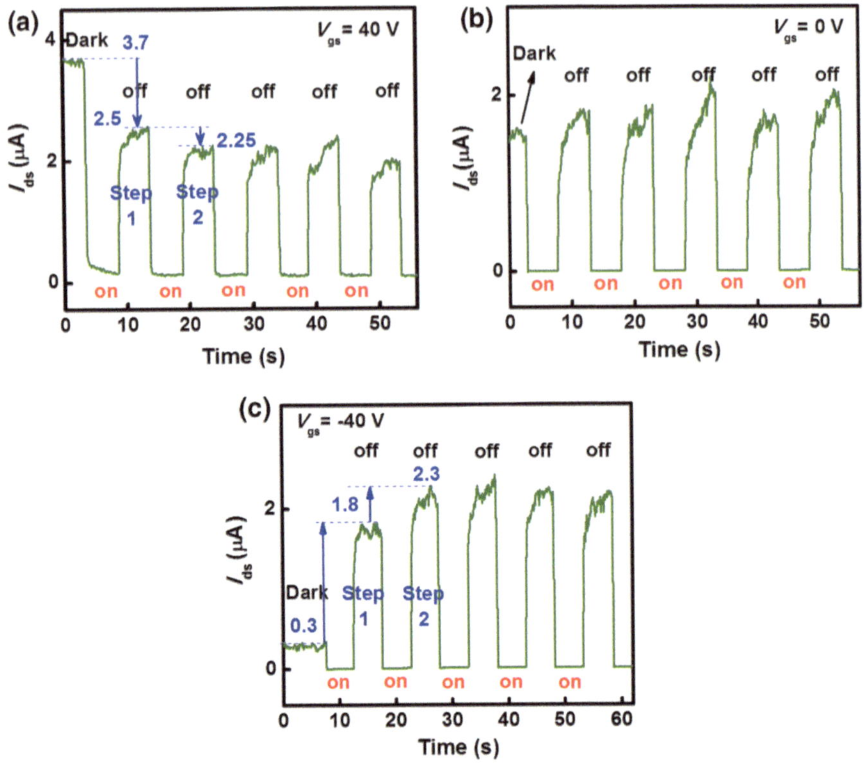


**Fig. 3.5** Temporal response of InAs NW transistor (532 nm, 8 mW mm$^{-2}$) at $V_{ds}$ = 1 V in air, acquired for gate voltage of 40 V (**a**), 0 V (**b**) and −40 V (**c**). The photoconductance modulation is induced by operating a shutter for the light on/off at an interval of 5 s

alternatively with an interval of 5 s. As presented in Fig. 3.5a, the light modulation leads to a "low" current state under illuminated conditions, ($I_{on}$), and a "high" current state under dark conditions, ($I_{off}$). Highly stable and reproducible characteristics are shown by the anomalous switching between these two states. It is noteworthy that in Fig. 3.5a, $I_{Dark} \approx 3.7$ μA is found at $V_{gs} = 40$ V before illumination, but after the first 5 s light exposure, $I_{off\text{-}step1}$ decreases to 2.5 μA, and after the second 5 s light exposure, $I_{off\text{-}step2}$ decreases to 2.25 μA. After that, $I_{off}$ remains relatively stable ($I_{off}/I_{on} \approx 23$) with a small fluctuation during the on-off cycles of light. The response following the light on-off cycles with $V_{gs} = 0$ V is given in Fig. 3.5b. $I_{off} \approx I_{Dark}$ and $I_{off}/I_{on} \approx 133$ can be obtained in that response. When $V_{gs} = -40$ V (see Fig. 3.5c), the curve exhibits an opposite trend of change compared with the case in Fig. 3.5a. A reduced free electron density in the channel is induced by the negative gate voltage, $I_{Dark} \approx 0.3$ μA. $I_{off\text{-}step1}$ increases to 1.8 μA after the first 5 s light exposure, and $I_{off\text{-}step2}$ increases to 2.3 μA after the second 5 s light exposure. After that, $I_{off}$ remains at a stable high level during the on-off cycles of light compared with $I_{Dark}$, and $I_{off}/I_{on}$ is found to be as high as 115.

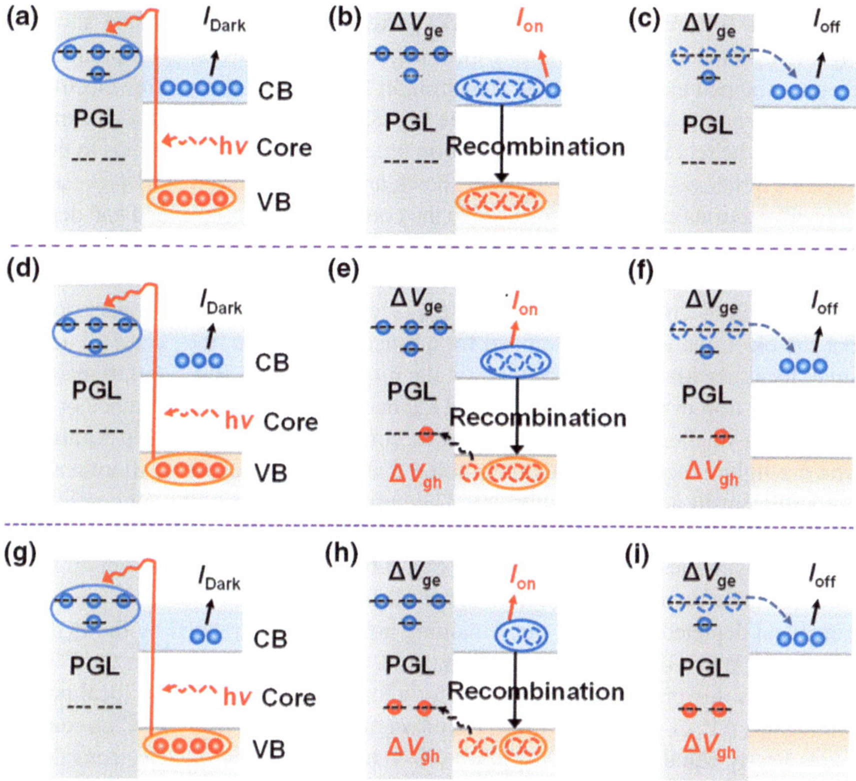


**Fig. 3.6** Schematic of the energy band diagrams of the InAs NW to demonstrate the photoresponse mechanism. CB and VB are the conduction and valence band, respectively. $\Delta V_{ge}$ represents the negative photogate and $\Delta V_{gh}$ represents the positive photogate. The blue spheres are electrons and the red spheres are holes

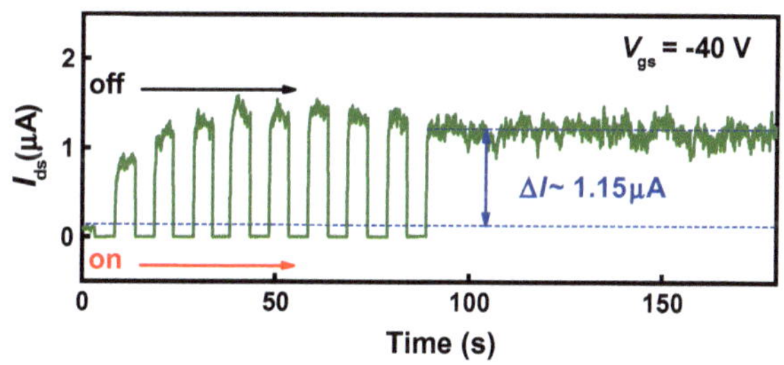


**Fig. 3.7** Temporal response of another InAs NW transistor (532 nm, 8 mW mm$^{-2}$) at $V_{ds} = 1$ V and $V_{gs} = -40$ V in air

The photoresponse mechanism of NWs with PGLs based on the experimental observation mentioned above is schematically illustrated in Fig. 3.6. (1) Photogenerated electrons from the core, upon light exposure with photon energy higher than NW bandgap, are excited into the PGL (see Fig. 3.6a). All photoexcited holes are immediately recombined by the majority of free electrons in the core, which are responsible for conductivity in the dark, corresponding to the situation shown in Fig. 3.5a. At the same time, a strong local negative photogate ($\Delta V_{ge}$) is produced by keeping trapped electrons in PGL, resulting in further depletion of free electrons in the core through capacitive coupling (see Fig. 3.6b). The rapid decrease of current is caused by the recombination and depletion processes, which corresponds to the fast response process. Actually, Fig. 3.6b shows an equilibrium physical process. The trapped electrons can also be released to the core, and then recombined and depleted again with light illumination. $I_{on}$ is the result of the recombination-depletion-release cycle. When the light is blocked, trapped electrons are released immediately from the PGL with decreasing $\Delta V_{ge}$ to form $I_{off}$ (see Fig. 3.6c), i.e. the recovery process. For the electrons which are trapped for a much longer time, they can still act as a small local negative photogate $\Delta V_{ge}$ on the n-type NW (see Fig. 3.6c), remaining a weak depletion in the core channel, and $I_{off}$ decreases. Thus, the existence of $\Delta V_{ge}$ results in a smaller $I_{off\text{-}step1}$ (2.5 μA) when compared with $I_{Dark}$ (3.7 μA) after the first 5 s light exposure. It should be noted that, in this case, 5 s of illumination is not sufficient to achieve equilibrium. The current $I_{off\text{-}step2}$ (2.25 μA) continues to decrease, following $I_{off\text{-}step1}$ until the current reaches a stable state. (2) The number of electrons in the core is reduced (see Fig. 3.6d), as $V_{gs}$ decreases from 40 to 0 V (refer to the situation in Fig. 3.5b). After all free electrons in the core are recombined and depleted ($I_{on} \approx 0$), the remaining holes are also trapped by the PGL (see Fig. 3.6e). When the light is blocked, most of the trapped electrons in the PGL are released. In this process, holes trapped for a long time which act as a local positive photogate ($\Delta V_{gh}$), can almost offset the effect of $\Delta V_{ge}$ (see Fig. 3.6f). Consequently, $I_{off}$ is nearly equal to $I_{Dark}$. (3) $V_{gs} = -40$ V further reduces free electrons density in the core (see Fig. 3.6g), corresponding to the situation shown in Fig. 3.5c. As a result, only a small proportion of photogenerated holes is needed for the recombination process and the rest of them are trapped by the PGL (see Fig. 3.6h). The photogenerated electrons accumulated in the PGL during the first 5 s light exposure are released to the core after the light is blocked, leading to an increasing $I_{off\text{-}step1}$. In the mean time, since $\Delta V_{gh}$ caused by the accumulated holes in the PGL is much higher than $\Delta V_{ge}$ (see Fig. 3.6i), the strong positive $\Delta V_{gh}$ enhances electron density in the core. Similar to the case mentioned in (1), in order to reach equilibrium, 5 s of illumination is not sufficient, so this process will continue until $I_{off}$ is stable. Therefore, $I_{off\text{-}step2}$ (2.3 μA) keeps increasing following $I_{off\text{-}step1}$ (1.8 μA) until the current reaches a stable state. It is worth noting that the longtime trapped photogenerated electrons and holes are hard to be released or recombined in Fig. 3.6c, f, i. It is found that for NW, it takes at least one hour to recover to its original state in the dark. Therefore, the photogating effect can last for a long time, and InAs NW transistors can be used in both sensitive photodetectors and light-induced current amplifiers under the negative $V_{gs}$ modulation. The temporal response of another InAs NW transistor at $V_{gs} = -40$ V is given in Fig. 3.7, which obviously shows that the enhancement of $I_{off}$ can last for a long time. As a matter of fact, $\Delta V_{gh}$ can also offset a fraction

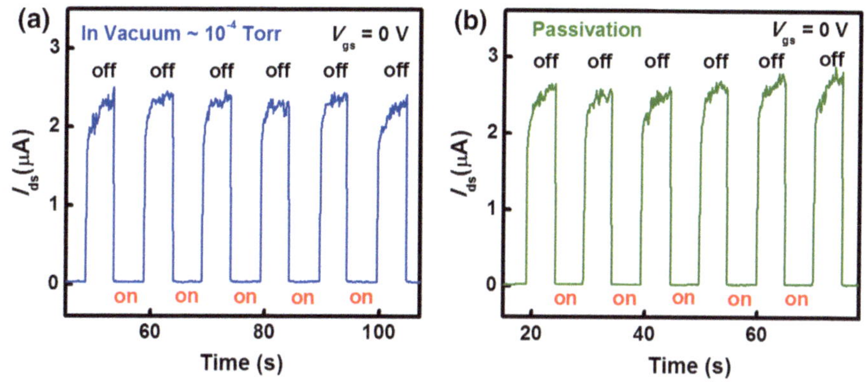


**Fig. 3.8** Photoresponse properties of InAs NW transistor (532 nm, 8 mW mm$^{-2}$) at $V_{gs} = 0$ V and $V_{ds} = 1$ V in vacuum (**a**) and with a 10 nm $HfO_2$ passivation layer (**b**). Atomic layer deposition technique was used to deposit the 10 nm $HfO_2$ dielectric

of the negative $\Delta V_{ge}$ during the light exposure process (refer to Fig. 3.6e, h). In our experiment, the negative photogate is the dominating factor due to the accumulation of more photoexcited electrons than holes in the PGL, and the effect of depletion in the channel is still strong. That is why $I_{on}$ can remain a stable low current at different $V_{gs}$.

To explain the effect of surface states on anomalous photoresponse, the response properties of the device in vacuum condition and with a 10 nm $HfO_2$ passivation layer were studied, results of which are shown in Fig. 3.8. It can be seen that the device works well in both air and vacuum. What is more, the device performance is not affected by the passivation layer. Photodetectors with positive photoresponse made of single-crystalline InAs NW grown by molecular beam epitaxy have been reported, in which photocurrent was significantly influenced by gas molecules and passivation layer [22]. Chemical molecules adsorbed on the surface of InAs NW could alter the band alignment to weaken the strength of built-in electric field of NW-metal junction by way of electrostatic interaction, and they could also transfer photocarriers through surface defect states to reduce photocurrents. In this case, the $HfO_2$ passivation layer could protect the NW from ambient environment and surface defect states so that the device performance could be improved. But in our work, vacuum testing condition and passivation layer had no effect on the anomalous photoresponse. So, the effect of surface states is insignificant compared with that of photogating layer.

The response speed as a key parameter determines the capability of a photodetector to follow a fast-varying optical signal [20]. A single normalized modulation cycle measured with a light intensity of 8 mW mm$^{-2}$ at $V_{gs} = 0$ V and $V_{ds} = 1$ V is given in Fig. 3.9. In this case, the response time (falling time $t_f$), defined as the time necessary for the current to decrease from 90% $I_{peak}$ to 10% $I_{peak}$, is found to be 12 ms, and the recovery time (rising time $t_r$), defined homoplastically, is found to be 3.92 s. Moreover, two time constants (fitting the rising edge by equation $I = I_0\,[1 - A\exp(-t/\tau_{rec1}) - B\exp(-t/\tau_{rec2})]$), $\tau_{rec1}$ (140 ms) and $\tau_{rec2}$ (3.12 s), are found that

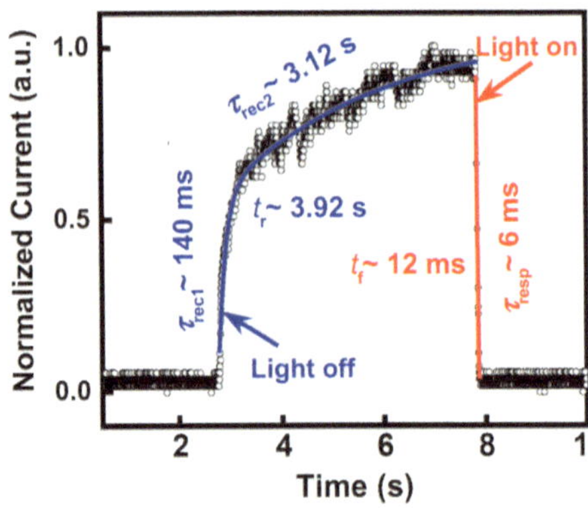


**Fig. 3.9** A single normalized modulation cycle measured with light intensity 8 mW mm$^{-2}$ (open circles) at $V_{gs}=0$ V and $V_{ds}=1$ V. An oscilloscope was used to monitor the time dependence of the current. The blue and green fitting lines correspond to the rising (recovery $\tau_{rec1}$ and $\tau_{rec2}$) and falling (response $\tau_{resp}$) edges, respectively

correspond to fast and slow recovery processes, respectively. Another time constant of $\tau_{resp}$ (6 ms) (fitting the falling edge by equation $I = I_0\ [1 + C\ \exp(-t/\tau_{resp})]$) is also obtained for the fast response process. The release process of the trapped photoexcited electrons from the PGL and the decay of photogating effect after the light was blocked are indicated by $\tau_{rec1}$ and $\tau_{rec2}$ in the recovery process, and the fast recombination process and the enhancement of photogating effect with light excitation are indicated by $\tau_{resp}$.

To investigate the sensitivity of phototransistor, photoresponse measurements with different light intensities were performed and the results are shown in Fig. 3.10a. In it, $I_{off}/I_{on} \approx 133$, 125 and 30 can be obtained with the light intensities of 8, 4 and 0.8 mW mm$^{-2}$, respectively, at $V_{gs}=0$ V and $V_{ds}=1$ V; $I_{off}$ and $I_{on}$ decreases and increases with the decay of light power, accordingly. The reason is that fewer electron-hole pairs are generated at lower intensities of incident light, so that more electrons stay in the core during the recombination and depletion process, helping to increase $I_{on}$. At the same time, the limited number of holes trapped in the PGL results in a weakened $\Delta V_{gh}$. Therefore, the negative $\Delta V_{ge}$ is dominant and reduces $I_{off}$. As presented in Fig. 3.10b, the relation between $I_{off}$ and light intensity obeys the power law [13, 15, 20, 42] $I = cP^k$, where $P$ is the light intensity, $c$ is a proportionality constant, and $k$ is an empirical value. By fitting the measured data, $c=1$ and $k = 0.35$ were obtained. This relation further indicates that accumulated photoexcited electrons in the PGL indeed contribute to $I_{off}$. The non-integer power law ($k < 1$) suggests there is a complex carrier transfer process shown in Fig. 3.6d–f.

Based on the above discussion, it is clear that the junction region has no role in producing photocurrents. In order to prove this point, a long InAs NW was fabricated into a multi-electrode transistor to study the position dependence of photoelectric characteristics. In Fig. 3.11a, 3 metallic electrodes (2 μm wide) are placed in contact with the NW. Three channels, i.e. CH1, CH2 and CH3, are formed along the NW with length of 3, 1.5 and 6.5 μm, respectively. $I_{ds}$-$V_{ds}$ curves of the as-fabricated InAs NW transistor in the dark and under 532 nm illumination in air are given in Fig. 3.11b. Relative magnitude of photocurrents is $I_{Light\text{-}CH1} > I_{Light\text{-}CH2} \approx I_{Light\text{-}CH3}$. The photocurrent of CH3 is completely determined by that of CH2. That is, the resistance

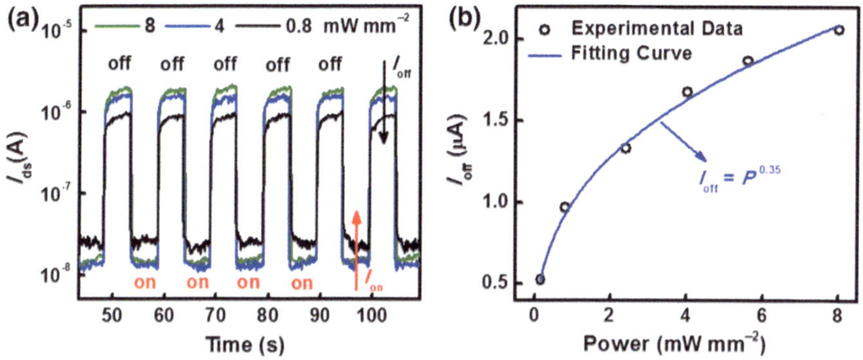


**Fig. 3.10** Power dependence of photoresponse. **a** Photoresponse properties of the InAs NW transistor with 532 nm illumination at $V_{gs} = 0$ V and $V_{ds} = 1$ V in air, acquired for different light intensities: 0.8, 4 and 8 mW mm$^{-2}$. By manipulating a shutter for the light on-off at an interval of 5 s, the photoconductance modulation is induced. **b** $I_{off}$ versus light intensity

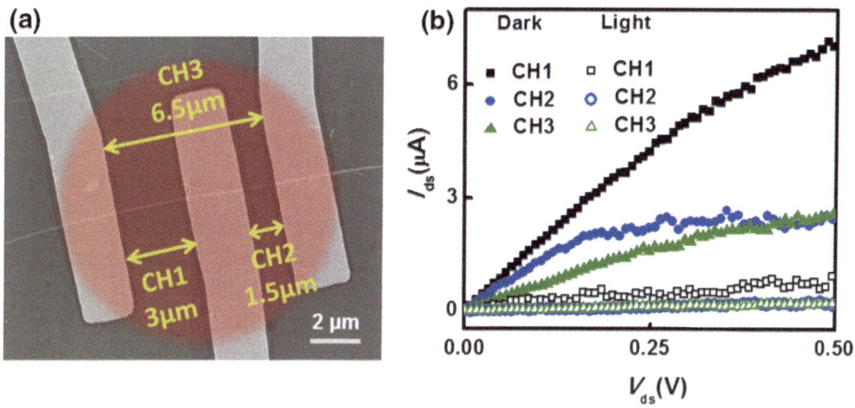


**Fig. 3.11** Position dependence of photoresponse and near-infrared measurement. **a** SEM image of the InAs NW transistor. **b** $I_{ds}$-$V_{ds}$ curves of the multi-electrode InAs NW transistor in the dark (closed symbols) and under light illumination (532 nm, 8 mW mm$^{-2}$) (open symbols) at $V_{gs} = 0$ V in air recorded for three conductive channels: CH1, CH2 and CH3

of CH3 is equal to the part of the NW under light excitation which has the greatest resistance, i.e. CH2 in this case. Therefore, high performance of photodetectors can be easily obtained as long as any small segment is sensitive.

The performance of the InAs NW transistor under near-infrared light illumination was studied at room temperature. Figure 3.12 shows the $I_{ds}$-$V_{ds}$ properties obtained under light illumination with a wavelength of 1.2 μm. It is obvious that the photocurrents decrease evidently on account of the photogating effect mentioned above. Photoconductive gain of −1.1 is acquired at $V_{ds} = 2$ V and the reduplicate measurements prove that the device has a desirable stability. Besides, it is clear that the device produces a higher negative response to visible light than to near-infrared light, as in

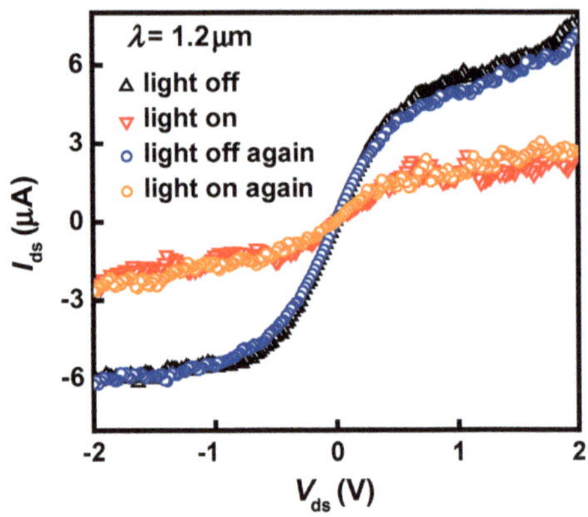


**Fig. 3.12** $I_{ds}$-$V_{ds}$ curves of the InAs NW transistor under near-infrared light excitation at $V_{gs} = 0$ V in air

Fig. 3.11b. The reason may be that high energy photons excite the photocarriers into the PGL more efficiently, resulting in a stronger negative photogate to modulate the channel conductance.

## 3.3 Conclusions

High photoconductive gain of ~$-10^5$ and fast response time of 12 ms are achieved in InAs NW phototransistors at room temperature. Further research shows that the good device performance can be attributed to the trapping mechanism of self-assembled near-surface PGL, which results in a strong photogating effect on the NW channel. In addition, the device could readily find much wide range of applications because its photodetection is free from environment impact. The new phenomenon, majority carriers contributing the photocurrents, together with related properties, will pave a way to realize novel high-sensitive broadband spectrum room-temperature detection.

## References

1. K. Takei, T. Takahashi, J.C. Ho, H. Ko, A.G. Gillies, P.W. Leu, R.S. Fearing, A. Javey, Nat. Mater. **9**, 821 (2010)
2. D. Saxena, S. Mokkapati, P. Parkinson, N. Jiang, Q. Gao, H.H. Tan, C. Jagadish, Nat. Photonics **7**, 963 (2013)
3. M.S. Gudiksen, L.J. Lauhon, J. Wang, D.C. Smith, C.M. Lieber, Nature **415**, 617 (2002)
4. Y. Huang, X. Duan, Y. Cui, L.J. Lauhon, K.-H. Kim, C.M. Lieber, Science **294**, 1313 (2001)
5. J. Wallentin, N. Anttu, D. Asoli, M. Huffman, I. Åberg, M.H. Magnusson, G. Siefer, P. Fuss-Kailuweit, F. Dimroth, B. Witzigmann, H.Q. Xu, L. Samuelson, K. Deppert, M.T. Borgström, Science **339**, 1057 (2013)
6. J. Tang, Z. Huo, S. Brittman, H. Gao, P. Yang, Nat. Nanotechnol. **6**, 568 (2011)

7. S. Ju, A. Facchetti, Y. Xuan, J. Liu, F. Ishikawa, P. Ye, C. Zhou, T.J. Marks, D.B. Janes, Nat. Nanotechnol. **2**, 378 (2007)
8. D.C. Dillen, K. Kim, E.-S. Liu, E. Tutuc, Nat. Nanotechnol. **9**, 116 (2014)
9. T. Takahashi, K. Takei, E. Adabi, Z. Fan, A.M. Niknejad, A. Javey, ACS Nano **4**, 5855 (2010)
10. A.C. Ford, J.C. Ho, Y.-L. Chueh, Y.-C. Tseng, Z. Fan, J. Guo, J. Bokor, A. Javey, Nano Lett. **9**, 360 (2009)
11. N. Han, F. Wang, J.J. Hou, S.P. Yip, H. Lin, F. Xiu, M. Fang, Z. Yang, X. Shi, G. Dong, T.F. Hung, J.C. Ho, Adv. Mater. **25**, 4445 (2013)
12. S.A. Dayeh, D.P.R. Aplin, X. Zhou, P.K.L. Yu, E.T. Yu, D. Wang, Small **3**, 326 (2007)
13. H. Kind, H. Yan, B. Messer, M. Law, P. Yang, Adv. Mater. **14**, 158 (2002)
14. J.-J. Wang, F.-F. Cao, L. Jiang, Y.-G. Guo, W.-P. Hu, L.-J. Wan, J. Am. Chem. Soc. **131**, 15602 (2009)
15. S.-C. Kung, W.E. van der Veer, F. Yang, K.C. Donavan, R.M. Penner, Nano Lett. **10**, 1481 (2010)
16. Y.H. Ahn, J. Park, Appl. Phys. Lett. **91**, 162102 (2007)
17. Z. Wang, M. Safdar, C. Jiang, J. He, Nano Lett. **12**, 4715 (2012)
18. G. Chen, B. Liang, X. Liu, Z. Liu, G. Yu, X. Xie, T. Luo, D. Chen, M. Zhu, G. Shen, Z. Fan, ACS Nano **8**, 787 (2014)
19. Z. Liu, G. Chen, B. Liang, G. Yu, H. Huang, D. Chen, G. Shen, Opt. Express **21**, 7799 (2013)
20. P. Wu, Y. Dai, Y. Ye, Y. Yin, L. Dai, J. Mater. Chem. **21**, 2563 (2011)
21. C. Soci, A. Zhang, B. Xiang, S.A. Dayeh, D.P.R. Aplin, J. Park, X.Y. Bao, Y.H. Lo, D. Wang, Nano Lett. **7**, 1003 (2007)
22. J. Miao, W. Hu, N. Guo, Z. Lu, X. Zou, L. Liao, S. Shi, P. Chen, Z. Fan, J.C. Ho, T.-X. Li, X.S. Chen, W. Lu, ACS Nano **8**, 3628 (2014)
23. P. Wu, Y. Dai, T. Sun, Y. Ye, H. Meng, X. Fang, B. Yu, L. Dai, A.C.S. Appl, Mater. Interfaces **3**, 1859 (2011)
24. Q. Hang, F. Wang, P.D. Carpenter, D. Zemlyanov, D. Zakharov, E.A. Stach, W.E. Buhro, D.B. Janes, Nano Lett. **8**, 49 (2008)
25. Y. Ahn, J. Dunning, J. Park, Nano Lett. **5**, 1367 (2005)
26. S. Thunich, L. Prechtel, D. Spirkoska, G. Abstreiter, A. Fontcuberta i Morral, A.W. Holleitner, Appl. Phys. Lett. **95**, 083111 (2009)
27. J. Zhou, Y. Gu, Y. Hu, W. Mai, P.-H. Yeh, G. Bao, A.K. Sood, D.L. Polla, Z.L. Wang, Appl. Phys. Lett. **94**, 191103 (2009)
28. C.-H. Lin, T.-T. Chen, Y.-F. Chen, Opt. Express **16**, 16916 (2008)
29. M.-W. Chen, C.-Y. Chen, D.-H. Lien, Y. Ding, J.-H. He, Opt. Express **18**, 14836 (2010)
30. T.J. Kempa, B. Tian, D.R. Kim, J. Hu, X. Zheng, C.M. Lieber, Nano Lett. **8**, 3456 (2008)
31. M. Heurlin, P. Wickert, S. Fält, M.T. Borgström, K. Deppert, L. Samuelson, M.H. Magnusson, Nano Lett. **11**, 2028 (2011)
32. Y. Guo, Y. Zhang, H. Liu, S.-W. Lai, Y. Li, Y. Li, W. Hu, S. Wang, C.-M. Che, D. Zhu, J. Phys. Chem. Lett. **1**, 327 (2010)
33. W. Wei, X.-Y. Bao, C. Soci, Y. Ding, Z.-L. Wang, D. Wang, Nano Lett. **9**, 2926 (2009)
34. S.P. Mondal, S.K. Ray, Appl. Phys. Lett. **94**, 223119 (2009)
35. Z. Guo, D. Zhao, Y. Liu, D. Shen, J. Zhang, B. Li, Appl. Phys. Lett. **93**, 163501 (2008)
36. H. Zhou, G. Fang, L. Yuan, C. Wang, X. Yang, H. Huang, C. Zhou, X. Zhao, Appl. Phys. Lett. **94**, 013503 (2009)
37. G. Konstantatos, M. Badioli, L. Gaudreau, J. Osmond, M. Bernechea, F.P.G. de Arquer, F. Gatti, F.H.L. Koppens, Nat. Nanotechnol. **7**, 363 (2012)

38. C.-H. Liu, Y.-C. Chang, T.B. Norris, Z. Zhong, Nat. Nanotechnol. **9**, 273 (2014)
39. H.J. Joyce, Q. Gao, H.H. Tan, C. Jagadish, Y. Kim, X. Zhang, Y. Guo, J. Zou, Nano Lett. **7**, 921 (2007)
40. J.J. Hou, N. Han, F. Wang, F. Xiu, S. Yip, A.T. Hui, T. Hung, J.C. Ho, ACS Nano **6**, 3624 (2012)
41. W. Choi, M.Y. Cho, A. Konar, J.H. Lee, G.-B. Cha, S.C. Hong, S. Kim, J. Kim, D. Jena, J. Joo, S. Kim, Adv. Mater. **24**, 5832 (2012)
42. Y. Jiang, W.J. Zhang, J.S. Jie, X.M. Meng, X. Fan, S.-T. Lee, Adv. Funct. Mater. **17**, 1795 (2007)

# Chapter 4
# Hybrid $WSe_2$-$In_2O_3$ Phototransistors with Ultrahigh Detectivity

**Abstract** This chapter will introduce a hybrid phototransistor consisting of a single $In_2O_3$ nanowire as the channel material and a multilayer $WSe_2$ nanosheet as the decorating sensitizer for photodetection. By the unique design of energy band alignment of the channel and sensitizer materials, an enhanced photogating effect is realized in the depletion regime of the device, leading to a superior detectivity over broadband spectra.

**Keywords** $WSe_2$ nanosheet · $In_2O_3$ nanowire · Hybrid structure
Photogating effect · Detectivity

## 4.1 Introduction

In the past ten years, two-dimensional (2D) materials have shown their great potential as photosensitive elements for future high performance optoelectronic devices [1–21]. The distinctive properties like high charge carrier mobilities in graphene and thickness-dependent spectral coverage in transition metal dichalcogenides, augur well for high gain-bandwidth and a wide spectral response for photodetection [14–16, 22, 23]. However, the inherent weak optical absorption of the atomic-thickness 2D materials causes a low responsivity. A promising strategy has been adopted to use strongly light absorbing materials with opposite doping polarity to decorate the channels for enhancing light utilization efficiency [22–27]. Photocarriers were separated by the built-in field at the channel-sensitizer interface. The trapped carriers in the sensitizer produced a photogate, which increased the conductance of the channel to achieve a current amplification. Ultrahigh responsivity of ~$10^7$ A $W^{-1}$ in hybrid graphene-quantum dot (QD) phototransistor has been reported, where QDs and graphene serve as the sensitizer and the channel, respectively [22]. However, large dark currents due to the lack of bandgap in graphene lead to low sensitivity and high power consumption. To address this issue, researchers use low-dimensional channel candidates with a bandgap, such as $MoS_2$ and thin-film InGaZnO, to deplete the dark currents by applying appropriate back-gate voltages [25–27]. Several

 
N. Guo, *Infrared Photodetectors Based on Low-Dimensional Materials*,
Springer Theses, https://doi.org/10.1007/978-981-13-2838-1_4

orders of magnitude reduction in dark currents have been demonstrated in the depletion regimes of $MoS_2$-QDs and InGaZnO-QDs hybrid structures compared with graphene-QDs phototransistors, but the corresponding photocurrents were also reduced because of the reduction of built-in field induced by the applied back-gate electric field with opposite direction. Therefore, in spite of tremendous efforts, the challenge still remains for low-dimensional phototransistors to operate at the optimal point considering the trade-off between responsivity and dark current [23].

Here, a multilayer $WSe_2$-$In_2O_3$ nanowire (NW) hybrid structure for ultrahigh sensitivity photodetection was designed, where $WSe_2$ was used as the sensitizer that modulates the NW conductance. The $In_2O_3$ NW channel could be efficiently depleted with an external back-gate electric field, which intensifies the strength of the built-in field at the $WSe_2$-$In_2O_3$ interface to ensure a stronger photogating effect due to the unique design of energy band alignment. The phototransistor could operate in the depletion regimes with extremely low dark currents and high photocurrents.

## 4.2 Results and Discussions

### *4.2.1 Device Structure and Working Principle*

Figure 4.1a depicts the multilayer $WSe_2$-$In_2O_3$ NW phototransistor configuration. The n-type $In_2O_3$ NWs used in this work were grown by chemical vapor deposition (CVD) (see method). The device fabrication process can be briefly described as follows. First, we fabricate the $In_2O_3$ NW back-gated transistor. The as-grown $In_2O_3$ NWs were physically transferred onto the Si/$SiO_2$ (100 nm) substrate. The source/drain (S/D) electrodes (15 nm Cr and 50 nm Au) were prepared by electron-beam lithography, metallization, and lift-off process. Then, a multilayer $WSe_2$ was transferred onto the $In_2O_3$ NW channel using a micromanipulator (see method).

The working principle of the device is illustrated through the schematic and energy band diagram shown in Fig. 4.1b–d. Due to the work function mismatch between multilayer $WSe_2$ (bandgap of ~1.2 eV) [1, 28–30] and $In_2O_3$ NW (bandgap of ~2.75 eV) [31, 32], a built-in field is formed at the interface leading to a band bending. The conduction/valence band edge of $WSe_2$ bends upward and that of $In_2O_3$ NW downward. And, the difference in electron affinity forms a barrier at the junction. Upon light exposure, when the photon energy is below the bandgap of $In_2O_3$ but above that of $WSe_2$, the photocarriers are excited entirely in $WSe_2$. A positive photogate is formed by the accumulated holes at the $WSe_2$-$In_2O_3$ interface, which modulates the NW conductance through capacitive coupling. In order to fully deplete the $In_2O_3$ NW channel, it is necessary to apply a back-gate voltage lower than the threshold voltage on the device. The built-in field at the $WSe_2$-$In_2O_3$ interface is intensified by the back-gate electric field (their electric fields are in the same direction) to form a stronger band bending. The enhanced photogating effect in the depletion regime enables the detector to operate at extremely low dark current with high responsivity,

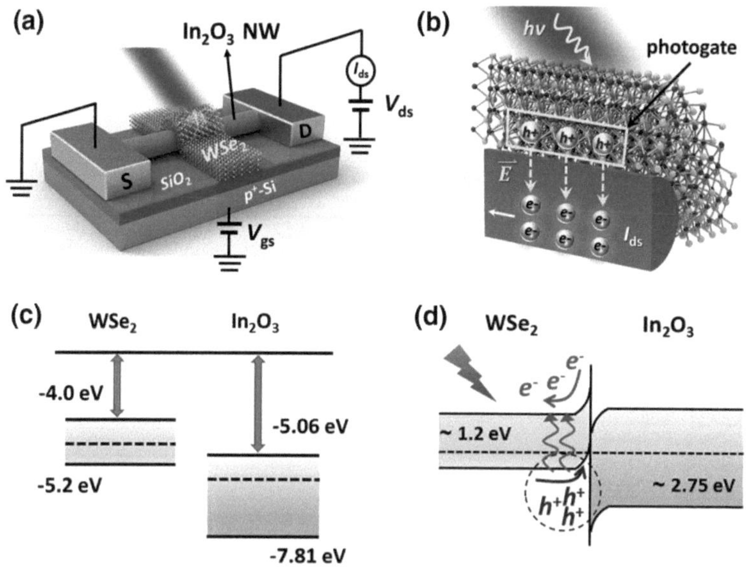


**Fig. 4.1** **a** A schematic illustration of the phototransistor, in which a multilayer $WSe_2$ is deposited onto the $In_2O_3$ NW channel. **b** Cross section of the device. The photoexcited holes in $WSe_2$ are accumulated at the interface of $WSe_2$ and $In_2O_3$ to modulate the NW conductance. **c** Energy band structure of the multilayer $WSe_2$ and $In_2O_3$ NW before they come into contact with each other. **d** Energy band diagram at the junction formed by $WSe_2$ and $In_2O_3$ NW

giving rise to a superior detectivity. It should be noted that Fig. 4.1d gives the equilibrium band diagram of the $WSe_2$-$In_2O_3$ junction in the depletion regime, where the Fermi level of n-type $In_2O_3$ NW is close to the middle of bandgap and that of $WSe_2$ is near the valence band edge due to the bipolarity of $WSe_2$. This will be discussed in more detail later.

Figure 4.2a, b shows an atomic force microscope (AFM) image of the hybrid structure and the height profile of $WSe_2$ and $In_2O_3$ NW. It can be seen that the NW height as measured from the uncovered portion is ~40 nm and the height of $WSe_2$ is ~80 nm. The 3D AFM image of the hybrid structure is given in Fig. 4.2c. The height profile of the overlapping area shows that only the top of $In_2O_3$ NW is in contact with $WSe_2$ (see Fig. 4.2d). In addition, the peaks at ~250 $cm^{-1}$ correspond to the first order $E_{2g}$ and $A_{1g}$ Raman modes suggesting that $WSe_2$ has a multilayer morphology (see Fig. 4.2e) [33–35]. The multilayer $WSe_2$ with a bandgap of ~1.2 eV can respond to light wavelength from visible to near infrared.

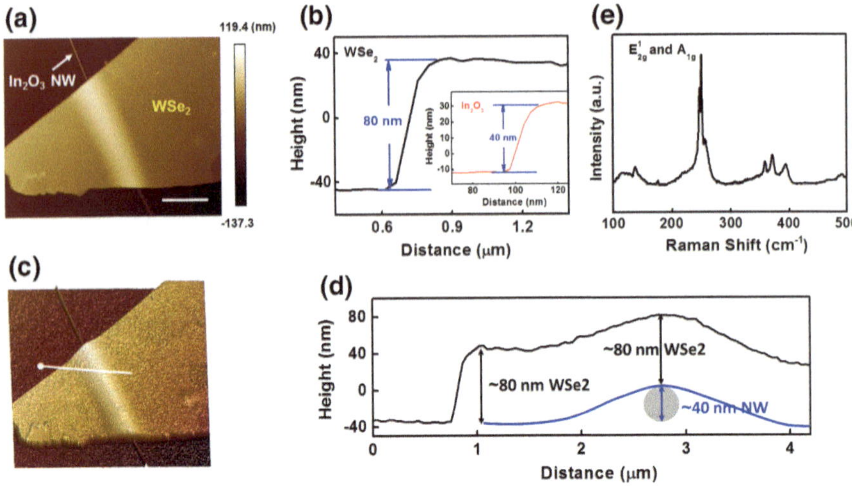


**Fig. 4.2** **a** AFM image of the hybrid structure. Scale bar, 2 μm. **b** Height profiles of $WSe_2$ (black line) and $In_2O_3$ NW (red line). **c** 3D AFM image of the hybrid structure. **d** Height profile along the white line in **c**. The blue line indicates the bottom of multilayer $WSe_2$ with a thickness of ~80 nm. The circle indicates the cross section of nanowire with a diameter of ~40 nm. **e** Raman spectrum of the multilayer $WSe_2$

## 4.2.2 Photoresponse Performance

In order to investigate the near infrared photoresponse, the optoelectric measurements of the phototransistor were performed under 940 nm light illumination. In Fig. 4.3a, the light illumination induces a distinct negative shift of the threshold voltage in transfer curves, indicating the existence of a positive photogate applied on the NW. So, a negative $V_{gs}$ is needed to offset the photogate. In addition, the higher light intensity can excite more photocarriers in $WSe_2$. The stronger photogate increases the NW conductance, resulting in a higher negative threshold voltage (marked by the dashed arrow). It should be noted that, benefiting from the large bandgap and efficient $V_{gs}$ modulation of the $In_2O_3$ NW, the detector can operate at full depletion. Moreover, the band bending at the interface becomes more pronounced as $V_{gs}$ changes from the accumulation to depletion regimes. An increased number of holes accumulated at the interface helps to make the photogate more efficient. A high photocurrent (~$10^{-7}$ A) with extremely low dark current ($10^{-14}$–$10^{-13}$ A) can be obtained in the depletion regime, which will be explained below in detail. For comparison with the hybrid structure, Fig. 4.3b gives $I_{ds}$-$V_{gs}$ curves of a bare $In_2O_3$ NW transistor under 940 nm illumination. It can be seen that there is no obvious change in the currents. Figure 4.3c presents the normalized spectral responsivity of the bare $In_2O_3$ NW transistor, and no response can be observed when the photon energy is below the bandgap of $In_2O_3$ NW (corresponding to the wavelength of 450 nm). Therefore, the photoresponse of the hybrid phototransistor for the near infrared can be attributed to the multilayer

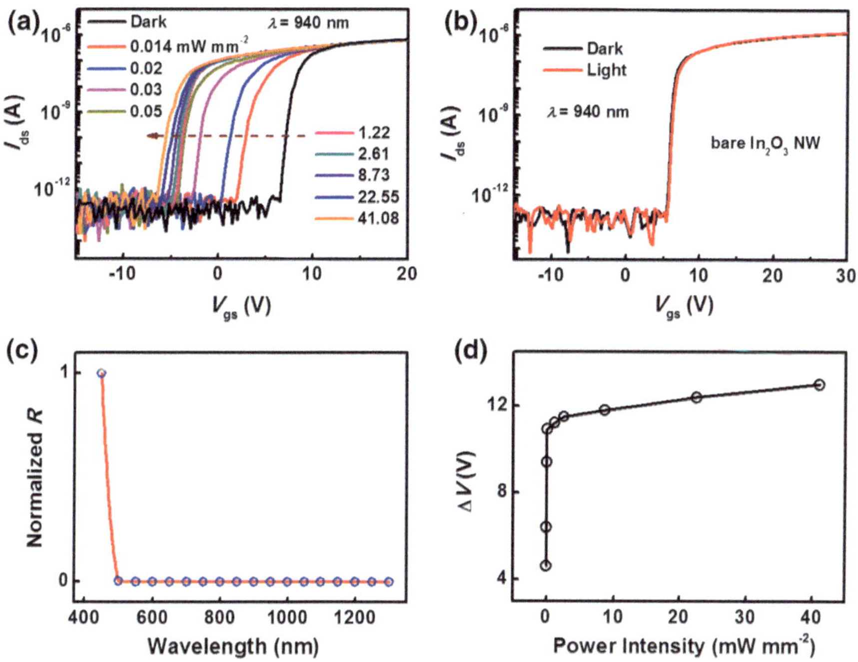


**Fig. 4.3** **a** $I_{ds}$-$V_{gs}$ curves of the phototransistor with different power intensities (940 nm) at $V_{ds} =$ 1 V. **b** $I_{ds}$-$V_{gs}$ curves of a bare $In_2O_3$ NW transistor in the dark and under 940 nm illumination. **c** Normalized spectral responsivity of the bare $In_2O_3$ NW transistor at $V_{ds} = 1$ V. **d** Shift of threshold voltage versus the light power intensity

$WSe_2$. Figure 4.3d shows the shift of threshold voltage ($\Delta V$) as a function of power intensity. A rapid rise of $\Delta V$ at low power intensity demonstrates the device's high photosensitivity.

The $I_{ds}$-$V_{ds}$ curves of phototransistor with different light power intensities are shown in Fig. 4.4a. The fact that $I_{ds}$ increases linearly with $V_{ds}$ proves a good ohmic contact between the Cr/Au electrodes and the NW. Under light illumination, a net photocurrent, defined as $I_{pc} = I_{light} - I_{dark}$, as high as ~113 nA is obtained at $V_{ds} =$ 1 V. The relation between $I_{pc}$ and light intensity ($P$) obeys the power law $I_{pc} \sim P^k$ (see Fig. 4.4b), where $k$ is an empirical value [36–38]. By fitting the measured data with the equation, $k = 0.018$ can be obtained. This relation further demonstrates that the accumulated photoexcited holes in $WSe_2$ are responsible for $I_{pc}$. Additionally, this $I_{pc}$ saturation with the increase of light intensity may be explained by the fact that no more photons can be absorbed by $WSe_2$ to excite excess carriers, forming a stronger photogate to further increase the NW conductance. The responsivity ($R$) as a function of light power intensity has been shown in Fig. 4.4c, given by $R = I_{ph}/PA$, where $I_{ph}$ is the photocurrent, $P$ is the power intensity, and $A$ is the effective photoactive area [39–42]. In our case, the photoactive part is the overlapped area of $WSe_2$ and $In_2O_3$ NW. Assuming that light incident on the effective portion is absorbed

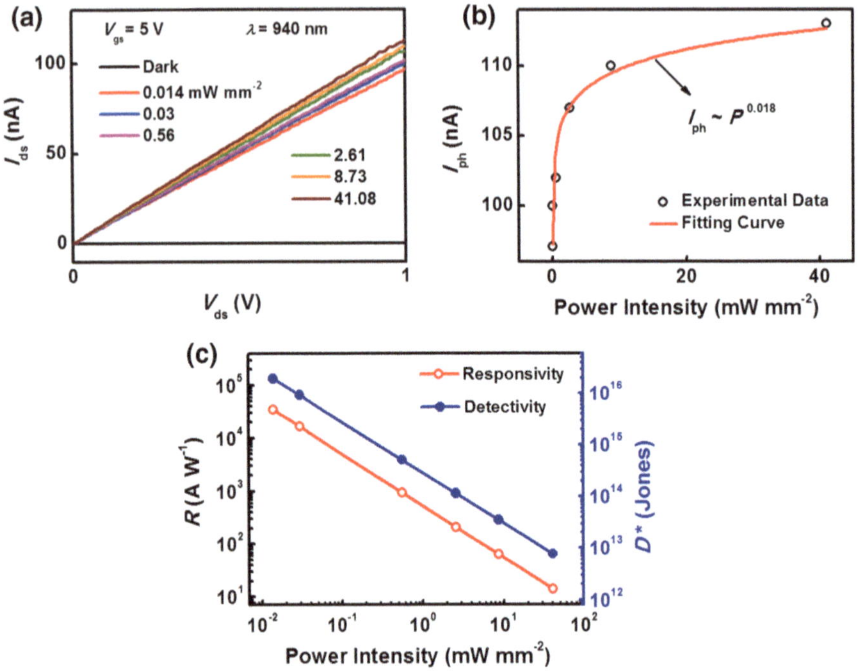


**Fig. 4.4** **a** $I_{ds}$-$V_{ds}$ curves of the phototransistor with different power intensities (940 nm). **b** Photocurrent versus light power intensity. **c** Responsivity and detectivity versus light power intensity

completely, a responsivity of $3.5 \times 10^4$ A $W^{-1}$ is obtained at $V_{ds} = 1$ V. The specific detectivity as an important parameter to determine the capability of a photodetector in responding to weak light signals is also shown in Fig. 4.4c. Assuming that shot noise is a major contributor of the dark current, the specific detectivity can be defined as $D^* = RA^{1/2}/(2eI_{dark})^{1/2}$, where $A$ is the effective photoactive area of the detector, $e$ is the electron charge, and $I_{dark}$ is the dark current [39, 43]. The measurement of noise current proves the assumption (please see next paragraph). The extremely low $I_{dark}$ ($10^{-14}$–$10^{-13}$ A) results in the maximum $D^*$ of $1.95 \times 10^{16}$ Jones. As expected in our design, this high performance is ascribed to the efficient photogating effect in the depletion regime.

Due to the extremely low dark current, the noise signal of the device cannot be captured by the test system. Figure 4.5 gives the background noise power density. If the background noise power density is equal to that from the detector, the detectivity $D^*$ can be obtained by:

$$D^* = \frac{\sqrt{A}\sqrt{B}}{\text{NEP}}, \quad \text{NEP} = \frac{\sqrt{\langle i_n \rangle^2}}{R}$$

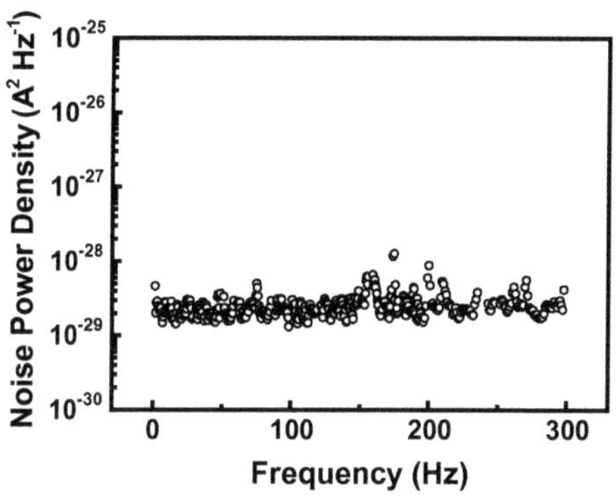


**Fig. 4.5** The background noise power density measurement

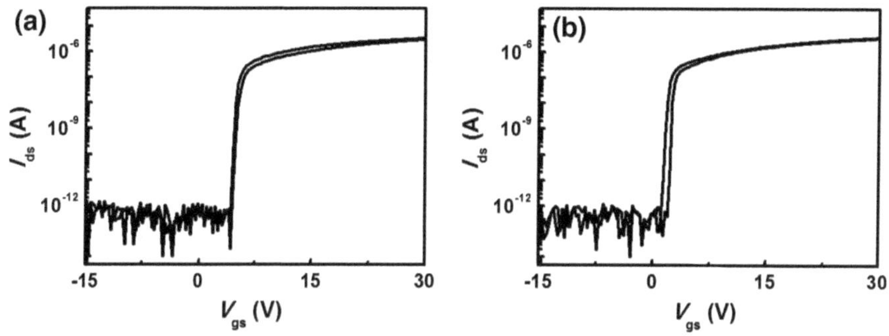


**Fig. 4.6** **a** and **b** are $I_{ds}$-$V_{gs}$ curves of another two bare $In_2O_3$ NW transistors in our work. No evident hysteresis exist in the $I_{ds}$-$V_{gs}$ curves

where $A$ is the effective photoactive area of the detector, $B$ is the bandwidth, $R$ is the responsivity, and $\langle i_n \rangle^2$ is the total noise current power which can be estimated by integrating the noise power density over the measured frequency range. Due to the constant noise power density ~$10^{-29}$ $A^2$ $Hz^{-1}$, the calculated $D^*$ of $4.9 \times 10^{14}$ Jones (1 Jones = 1 cm $Hz^{1/2}$ $W^{-1}$) is about two orders of magnitude smaller than the value calculated above $1.95 \times 10^{16}$ Jones in which the extremely low dark current is considered as the major contributor to the noise. The noise level of the detector is out of the scope of the system lower limit, which shows that the assumption mentioned above is valid. Figure 4.6 gives the transfer curves of the bare $In_2O_3$ NW back-gated transistors. The devices exhibit good electric properties and no evident hysteresis exist in transfer curve.

### 4.2.3 Temporal Response Characterization

To further elucidate the physical mechanism of phototransistor operation, its response speed has been studied. The photoconductance modulation is realized by switching the incident light on and off periodically. A high (low) current state under illumination (dark) condition, $I_{on}$ ($I_{off}$), has been obtained with light modulation (see Fig. 4.7a). The switching between the two states is highly stable and reversible. Figure 4.7b gives a single modulation cycle of temporal response. The rise edge consists of a fast component (current increases from 0 to 60% $I_{peak}$ (~32 ms)) and a slow component (current increases from 60% $I_{peak}$ to 100% $I_{peak}$ (~468 ms)). The former indicates the quick accumulation of photoexcited holes to form the photogate, and the latter indicates a slightly slower process to reach stable state with illumination. When the light was blocked, the current shows a sharp fall edge (~20 ms), attributable to the rapid carrier recombination process. In order to prove that the response speed of the device mainly depends on $WSe_2$, Fig. 4.7c gives the temporal response of a bare multilayer $WSe_2$ back-gated transistor under 940 nm illumination. Its response and

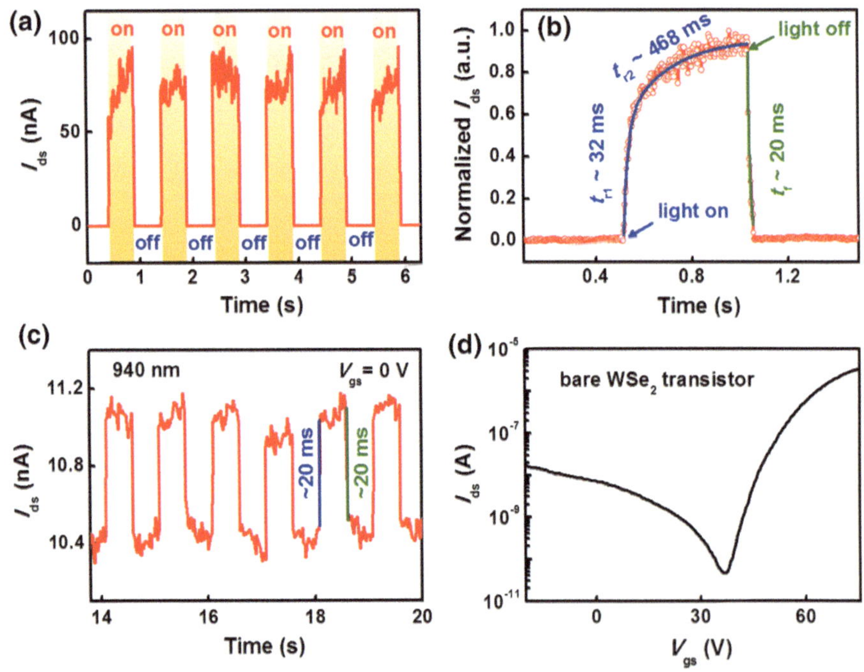


**Fig. 4.7** **a** Temporal response of the phototransistor (940 nm) at $V_{ds}$ = 1 V. The light is turned on/off at an interval of 1 s. **b** A single modulation cycle. An oscilloscope was used to monitor the time dependence of the current. The blue and green lines indicate the rising and falling edges, respectively. **c** Temporal response and **d** $I_{ds}$-$V_{gs}$ curve of the bare multilayer $WSe_2$ back-gated transistor

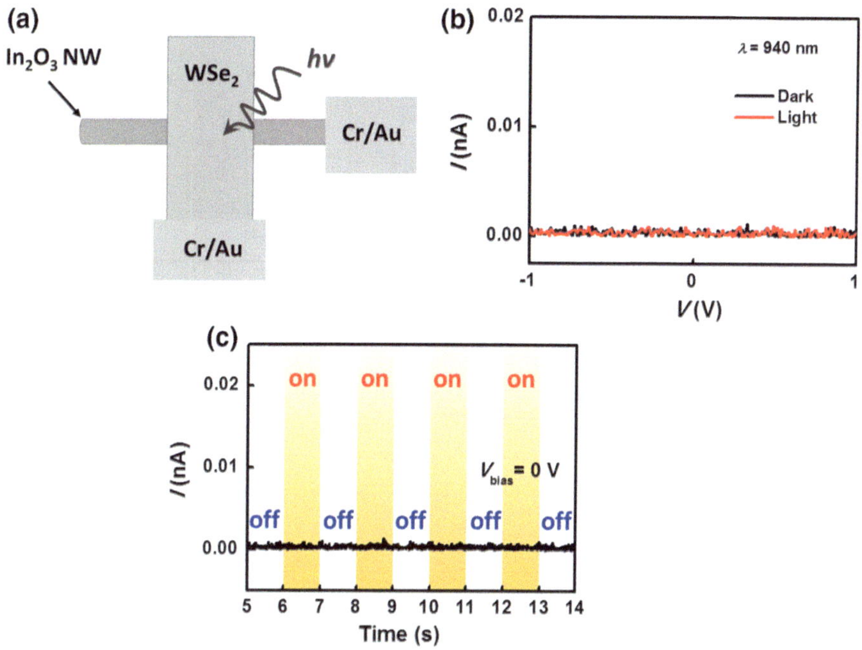


**Fig. 4.8** **a** A schematic illustration of the $WSe_2$-$In_2O_3$ heterostructure. **b** *I*-*V* curves of the heterostructure in the dark and under 940 nm illumination. **c** Temporal response of the heterostructure (940 nm)

recovery time are around 20 ms, which is basically consistent with the performance of $WSe_2$-$In_2O_3$ NW phototransistor.

Therefore, $WSe_2$ is believed to be responsible for response speed which can be improved if the photoactive material is characterized by fast carrier generation and recombination in the hybrid structure. Though an earlier work on $WSe_2$ phototransistor suggested that the response speed could be very fast (~5 μs) [44], in our case, the traps in the multilayer $WSe_2$ may have acted to lower the response speed. Moreover, it also can be seen in Fig. 4.7c that the photocurrent is less than 2 nA and the dark current is hard to be suppressed by applying gate bias due to the bipolar behaviors of $WSe_2$ (see Fig. 4.7d). The photodetector made solely of $WSe_2$ is difficult to achieve high responsivity and detectivity under near infrared light illumination with the same power intensity compared with hybrid phototransistor. The *I*-*V* property and temporal response of $WSe_2$-$In_2O_3$ heterostructure device under 940 nm light illumination are shown in Fig. 4.8. The fact that no photoresponse is observed proves that photoexcited carriers in $WSe_2$ do not transfer to $In_2O_3$ NW to help generate photocurrents.

Figure 4.9a–d give the photoresponse properties of a bare $In_2O_3$ NW transistor and hybrid phototransistor under 405 nm light illumination, respectively. The $In_2O_3$ NW transistor is characterized by a remarkable photocurrent and a slow response

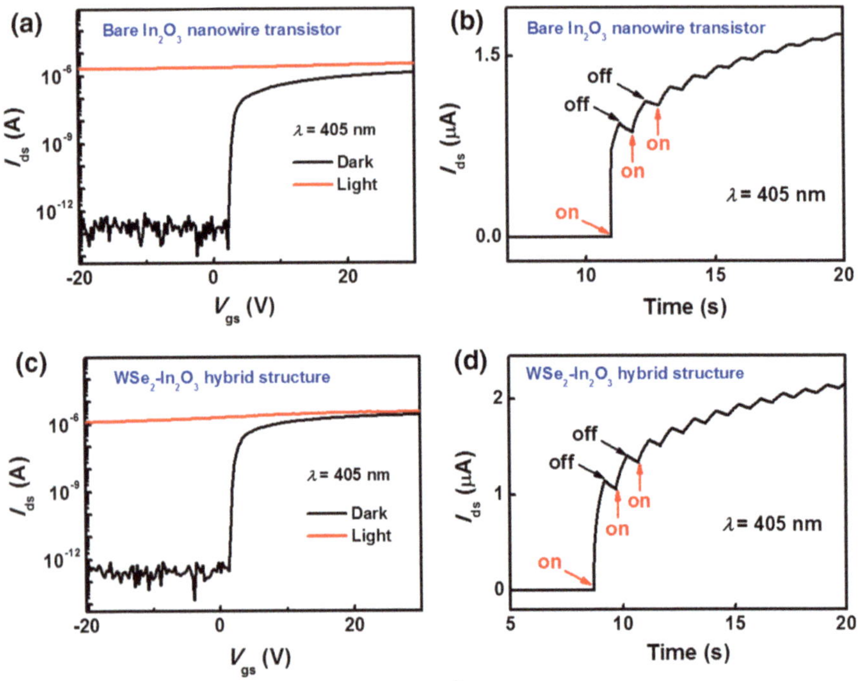


**Fig. 4.9** **a** $I_{ds}$-$V_{gs}$ and **b** temporal response curves of a bare $In_2O_3$ nanowire transistor in the dark and under 405 nm illumination. **c** $I_{ds}$-$V_{gs}$ and **d** temporal response curves of the hybrid $WSe_2$-$In_2O_3$ nanowire phototransistor in the dark and under 405 nm illumination. Power intensity, 11.51 mW mm$^{-2}$

speed due to its persistent photoconductivity, and the same it true of hybrid phototransistor. Therefore, when the photon energy is above the bandgap of $In_2O_3$ NW, the photocurrents are mainly generated in $In_2O_3$ NW.

### *4.2.4 Physical Mechanism*

Optoelectric properties of the device at 10 K were studied to investigate the photoresponse at low temperatures. In Fig. 4.10a, the transfer curves with different power intensities show that light illumination also leads to a distinct negative shift of the $V_{th}$ at low temperatures. With the increase of power intensity, the $V_{th}$ moves toward a larger negative $V_{gs}$, a similar trend observed at room temperature. Figure 4.10b depicts the $I_{ds}$-$V_{ds}$ characteristics with different power intensities. The photocurrents at 10 K are higher than that obtained at room temperature (see Fig. 4.4a) under the same illumination conditions. This is because the electron-phonon scattering is suppressed at low temperatures, leading to a higher mobility of $In_2O_3$ NW, in which case the carriers can be collected more efficiently by source-drain electrodes to form

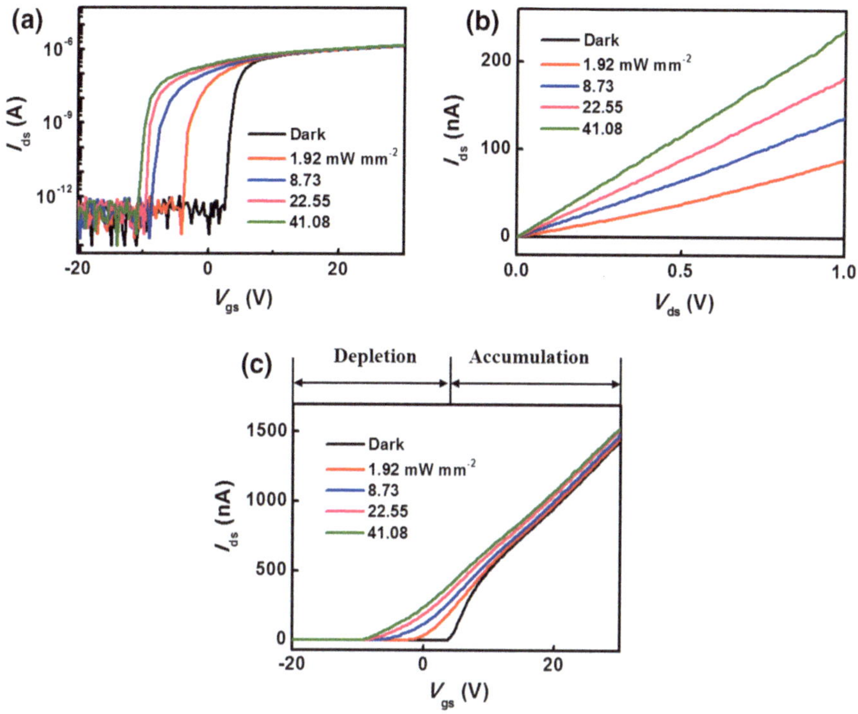


**Fig. 4.10** **a** $I_{ds}$-$V_{gs}$ and **b** $I_{ds}$-$V_{ds}$ curves of the phototransistor in the dark and under 940 nm illumination with different power intensities at 10 K. **c** Linear $I_{ds}$-$V_{gs}$ curves of the phototransistor in the dark and under 940 nm illumination with different power intensities at 10 K, corresponding to the semi-logarithmic curves in (**a**)

a high $I_{ds}$. Additionally, the low temperature provides a condition to study the effect of back gate on the strength of photogate without thermal influence. In Fig. 4.10c, semi-logarithmic curves in Fig. 4.1a are transformed into linear ones. Obviously, a larger photocurrent can be obtained in the depletion regime than in the accumulation regime.

Figure 4.11 shows the corresponding band alignment of the device in the accumulation and depletion regimes. When $V_{gs} = 30$ V, the position of Fermi level in multilayer $WSe_2$ is located close to the middle of bandgap because charge neutral point is at around 37 V (see Fig. 4.7d). And, the position of Fermi level in $In_2O_3$ NW is near the conduction band edge due to the n-type property. This little difference in work function (see the double-headed arrow in Fig. 4.11a) leads to a small band bending and fewer photoexcited holes accumulated at the interface, which generate a weak photogate and produce a smaller photocurrent. The case is quite different when $V_{gs} \leq V_{th}$, where the position of Fermi level in $WSe_2$ is near valence band edge and that in $In_2O_3$ NW is close to the middle of bandgap. The big difference in work function (see the double-headed arrow in Fig. 4.11b) leads to a large band bending and

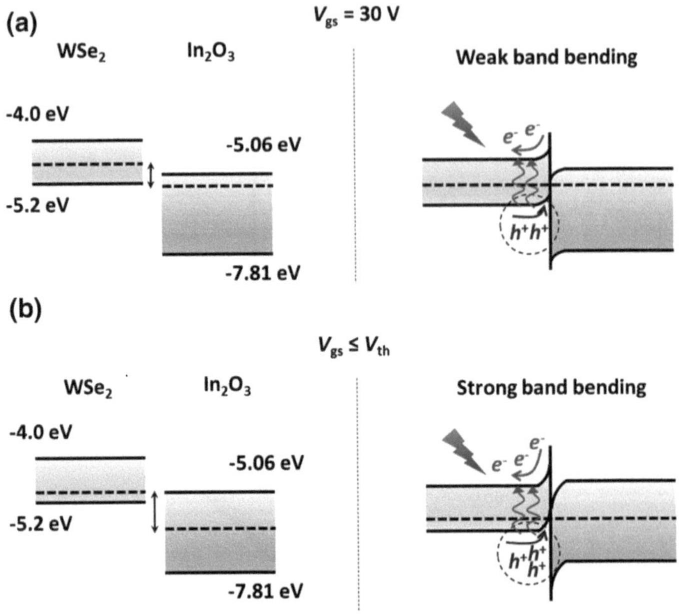


**Fig. 4.11** **a** The energy band diagram at the junction formed by $WSe_2$ and $In_2O_3$ when the detector operated in the accumulation regime. **b** The energy band diagram at the junction formed by $WSe_2$ and $In_2O_3$ when the detector operated in the depletion regime

more holes accumulated at the interface to generate a strong photogate, resulting in a large photocurrent. The advantage in our design is that the band bending becomes more pronounced from the accumulation to depletion regimes, enabling the device to operate at extremely low dark currents with high photocurrents.

Because of the bandgap of the multilayer $WSe_2$ ~ 1.2 eV, the hybrid phototransistor is expected to work from the visible to near infrared. Properties of broadband response of the device are investigated using a monochromator. In order to induce the response only in $WSe_2$, the wavelength ($\lambda$) varies from 550 to 1300 nm with an interval of 50 nm. The light intensity for different $\lambda$ remains consistent as 0.014 mW $mm^{-2}$. Figure 4.12a provides the partial data of $I_{ds}$-$V_{gs}$ characteristics with different $\lambda$ (another device in our work). It is obvious that the phototransistor responds to the light from visible to near infrared, and high photon energy leads to a more negative $V_{th}$. The incident photons with higher energy will excite the photocarriers in $WSe_2$ more efficiently to form a stronger photogate. That is the reason why $\Delta V$ increases with the decrease of $\lambda$ (see Fig. 4.12b). The spectral responsivities of the device have been shown in Fig. 4.12c. When $\lambda$ changes from visible to near infrared, the device's responsivity decreases by more than half from 550 to 750 nm. It remains relatively steady from 750 nm to 1.1 μm, and no response can be observed when the value of $\lambda$ is larger than 1.1 μm. It is noteworthy that the bandgap of multilayer $WSe_2$ is ~1.2 eV, corresponding to the $\lambda$ of ~1 μm. The extending cut-off wavelength may

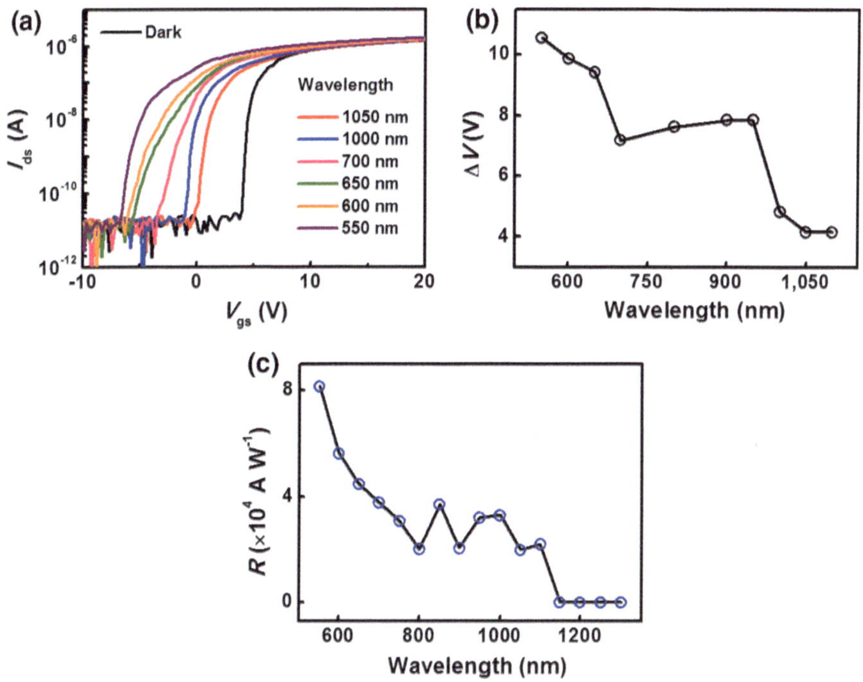


**Fig. 4.12** **a** Partial $I_{ds}$-$V_{gs}$ curves of the phototransistor in the dark and under light illumination with different wavelengths. **b** Shift of threshold voltage versus light wavelength. **c** Spectral responsivity of the phototransistor

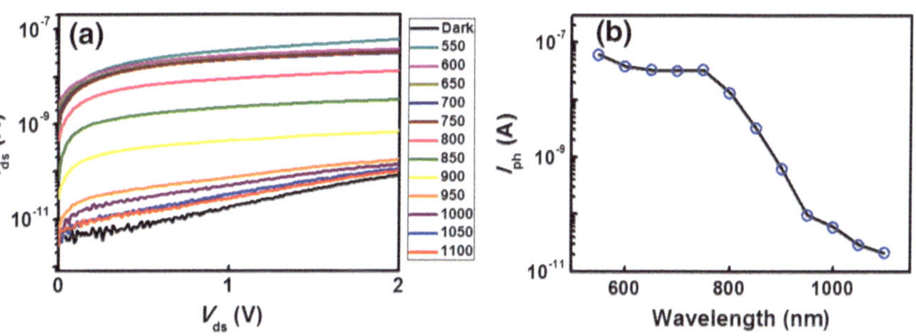


**Fig. 4.13** **a** $I_{ds}$-$V_{ds}$ curves of the bare multilayer $WSe_2$ back-gated transistor in the dark and under light illumination with different wavelengths. **b** Spectral response of the bare multilayer $WSe_2$ back-gated transistor

be ascribed to the defects in $WSe_2$. Figure 4.13 gives the spectral responsivity of a bare multilayer $WSe_2$ photodetector. Clearly, a smaller photocurrent is obtained in the near-infrared range up to 1.1 μm compared with the larger photocurrent in visible range. This demonstrates that it takes only a small number of photocarriers in $WSe_2$ to induce a high photoresponse in the hybrid structure.

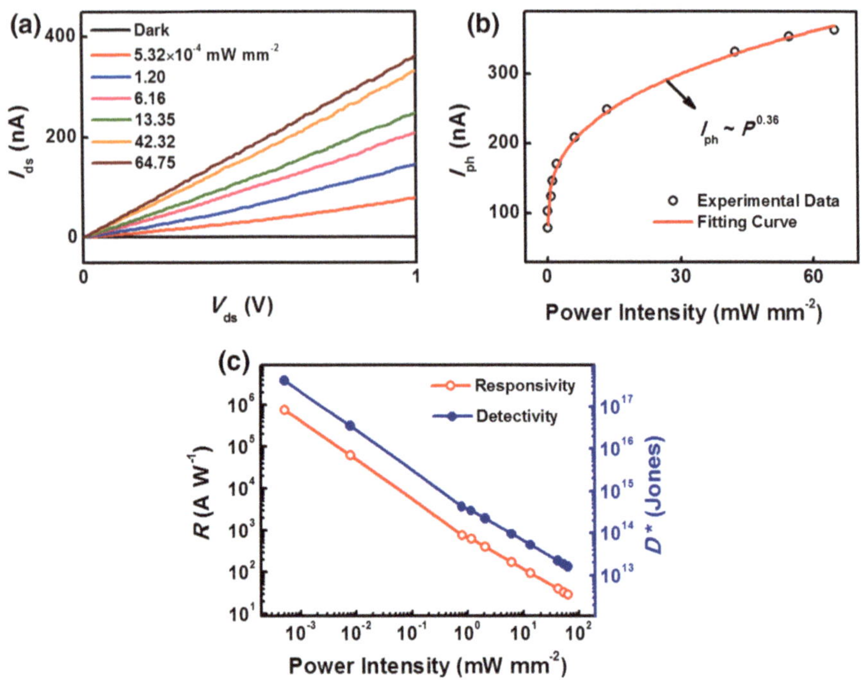


**Fig. 4.14** **a** $I_{ds}$-$V_{ds}$ curves of the phototransistor in the dark and under 637 nm illumination with different power intensities. **b** Photocurrent versus light power intensity. **c** Responsivity and detectivity versus light power intensity

To investigate photocurrent response for visible light, the optoelectric properties of the device were studied under 637 nm illumination. A net photocurrent $I_{pc}$ as high as ~364 nA is obtained at $V_{ds} = 1$ V (see Fig. 4.14a). Due to the high photon energy, a lower light intensity can induce a larger photocurrent in comparison with that shown in Fig. 4.4a. The relation between $I_{pc}$ and light intensity also obeys the power law $I_{pc} \sim P^k$ (see Fig. 4.14b). By fitting the measured data with the equation, $k = 0.36$ can be obtained. The maximum $R$ of $7.5 \times 10^5$ A W$^{-1}$ and $D*$ of $4.17 \times 10^{17}$ Jones at $V_{ds} = 1$ V are one order of magnitude larger than that measured under 940 nm illumination (see Fig. 4.14c). It shows that our device has extremely low dark current, ultrahigh $D*$, comparable $R$ and relatively fast response speed in the same operating condition.

## 4.3 Conclusions

We have designed a simple two-dimensional material and one dimensional nanowire hybrid architecture to achieve a broadband photoresponse with high performance.

The devices exhibit ultrahigh detectivity due to their enhanced photogating effect in the depletion regime, where extremely low dark currents and relatively high responsivity can be achieved at the same time. By decoupling the absorption from the charge transport with the former in $WSe_2$ and the later in $In_2O_3$, the phototransistors can respond to a wide spectral range from visible to near infrared. Our study provides a successful endeavor to optimize dark currents and responsivity for hybrid low-dimensional phototransistors with a unique design of energy band alignment of the channel and sensitizer materials.

## 4.4 Detailed Methods

**Nanowire Growth**: $In_2O_3$ NWs used in this study were prepared by CVD. $In_2O_3$ powder and graphite powder were mixed at a weight ratio of 10:1, which were then put into a quartz boat. Silicon substrate with pre-deposition of 1 nm thick of Au catalyst was placed upside-down atop the quartz boat. Then, the entire set-up was inserted into a tube reactor, heated to 1050 °C, and kept at this temperature for 1 h under a constant flow of gas (argon/oxygen = 100:1; 200 sccm). Finally, the substrate was left to be cooled to room temperature.

**Phototransistor Fabrication and Characterization**: The $WSe_2$ nanosheets were mechanically exfoliated on a $Si/SiO_2$ (285 nm) substrate. A thin layer of water-soluble polyvinyl acetate (PVA) was adhered to the poly-dimethyl siloxane (PDMS) to pick up the $WSe_2$ nanosheets. A micromanipulator was used to put the PDMS/PVA layer onto targets using microscope to locate the position. First, the PDMS/PVA layer picked up a $WSe_2$ nanosheet from the substrate, which was then precisely aligned to the target back-gated $In_2O_3$ NW transistor to form the hybrid structure. The PVA release from PDMS was performed through heating to soften the PVA layer. Finally, the PVA was removed from the hybrid structure in deionized water. The optoelectronic properties of the phototransistors were characterized using a Lake Shore Probe Station together with an Agilent B1500 semiconductor parameter analyzer. The spectral responsivity was performed using a supercontinuum spectrum laser source (400–2400 nm) combined with a monochromator and Agilent B2902.

## References


1. Q.H. Wang, K. Kalantar-Zadeh, A. Kis, J.N. Coleman, M.S. Strano, Nat. Nanotechnol. **7**, 699 (2012)
2. B.W.H. Baugher, H.O.H. Churchill, Y. Yang, P. Jarillo-Herrero, Nat. Nanotechnol. **9**, 262 (2014)
3. M. Massicotte, P. Schmidt, F. Vialla, K.G. Schädler, A. Reserbat-Plantey, K. Watanabe, T. Taniguchi, K.J. Tielrooij, F.H.L. Koppens, Nat. Nanotechnol. **11**, 42 (2016)
4. M. Long, E. Liu, P. Wang, A. Gao, H. Xia, W. Luo, B. Wang, J. Zeng, Y. Fu, K. Xu, W. Zhou, Y. Lv, S. Yao, M. Lu, Y. Chen, Z. Ni, Y. You, X. Zhang, S. Qin, Y. Shi, W. Hu, D. Xing, F. Miao, Nano Lett. **16**, 2254 (2016)

5. W.J. Yu, Y. Liu, H. Zhou, A. Yin, Z. Li, Y. Huang, X. Duan, Nat. Nanotechnol. **8**, 952 (2013)
6. S.-H. Jo, D.-H. Kang, J. Shim, J. Jeon, M.H. Jeon, G. Yoo, J. Kim, J. Lee, G.Y. Yeom, S. Lee, H.-Y. Yu, C. Choi, J.-H. Park, Adv. Mater. **28**, 4824 (2016)
7. D.-H. Kang, M.-S. Kim, J. Shim, J. Jeon, H.-Y. Park, W.-S. Jung, H.-Y. Yu, C.-H. Pang, S. Lee, J.-H. Park, Adv. Funct. Mater. **25**, 4219 (2015)
8. W. Zhang, M.-H. Chiu, C.-H. Chen, W. Chen, L.-J. Li, A.T.S. Wee, ACS Nano **8**, 8653 (2014)
9. O. Lopez-Sanchez, D. Lembke, M. Kayci, A. Radenovic, A. Kis, Nat. Nanotechnol. **8**, 497 (2013)
10. A. Pospischil, M. Humer, M.M. Furchi, D. Bachmann, R. Guider, T. Fromherz, T. Mueller, Nat. Photonics **7**, 892 (2013)
11. N.M. Gabor, J.C.W. Song, Q. Ma, N.L. Nair, T. Taychatanapat, K. Watanabe, T. Taniguchi, L.S. Levitov, P. Jarillo-Herrero, Science **334**, 648 (2011)
12. C.-H. Liu, Y.-C. Chang, T.B. Norris, Z. Zhong, Nat. Nanotechnol. **9**, 273 (2014)
13. N. Liu, H. Tian, G. Schwartz, J.B.-H. Tok, T.-L. Ren, Z. Bao, Nano Lett. **14**, 3702 (2014)
14. M. Liu, X. Yin, E. Ulin-Avila, B. Geng, T. Zentgraf, L. Ju, F. Wang, X. Zhang, Nature **474**, 64 (2011)
15. T. Mueller, F. Xia, P. Avouris, Nat. Photonics **4**, 297 (2010)
16. X. Wang, P. Wang, J. Wang, W. Hu, X. Zhou, N. Guo, H. Huang, S. Sun, H. Shen, T. Lin, M. Tang, L. Liao, A. Jiang, J. Sun, X. Meng, X. Chen, W. Lu, J. Chu, Adv. Mater. **27**, 6575 (2015)
17. J. Yao, Z. Zheng, G. Yang, Adv. Funct. Mater. **27**, 1701823 (2017)
18. J. Yao, Z. Zheng, J. Shao, G. Yang, A.C.S. Appl, Mater. Interfaces **7**, 26701 (2015)
19. J. Yao, Z. Zheng, G. Yang, A.C.S. Appl, Mater. Interfaces **8**, 12915 (2016)
20. J.D. Yao, Z.Q. Zheng, J.M. Shao, G.W. Yang, Nanoscale **7**, 14974 (2015)
21. J. Yao, J. Shao, Y. Wang, Z. Zhao, G. Yang, Nanoscale **7**, 12535 (2015)
22. G. Konstantatos, M. Badioli, L. Gaudreau, J. Osmond, M. Bernechea, F.P.G. de Arquer, F. Gatti, F.H.L. Koppens, Nat. Nanotechnol. **7**, 363 (2012)
23. D. Kufer, G. Konstantatos, ACS Photonics **3**, 2197 (2016)
24. Y. Liu, F. Wang, X. Wang, X. Wang, E. Flahaut, X. Liu, Y. Li, X. Wang, Y. Xu, Y. Shi, R. Zhang, Nat. Commun. **6**, 8589 (2015)
25. D. Kufer, I. Nikitskiy, T. Lasanta, G. Navickaite, F.H.L. Koppens, G. Konstantatos, Adv. Mater. **27**, 176 (2015)
26. D. Kufer, T. Lasanta, M. Bernechea, F.H.L. Koppens, G. Konstantatos, ACS Photonics **3**, 1324 (2016)
27. D.K. Hwang, Y.T. Lee, H.S. Lee, Y.J. Lee, S.H. Shokouh, J. Kyhm, J. Lee, H.H. Kim, T.-H. Yoo, S.H. Nam, D.I. Son, B.-K. Ju, M.-C. Park, J.D. Song, W.K. Choi, S. Im, NPG Asia Mater. **8**, e233 (2016)
28. A. Kumar, P.K. Ahluwalia, Eur. Phys. J. B **85**, 186 (2012)
29. P. Chen, T.T. Zhang, J. Zhang, J. Xiang, H. Yu, S. Wu, X. Lu, G. Wang, F. Wen, Z. Liu, R. Yang, D. Shi, G. Zhang, Nanoscale **8**, 3254 (2016)
30. B. Lei, Z. Hu, D. Xiang, J. Wang, G. Eda, C. Han, W. Chen, Nano Res. **10**, 1282 (2017)
31. D.O. Scanlon, A. Regoutz, R.G. Egdell, D.J. Morgan, G.W. Watson, Appl. Phys. Lett. **103**, 262108 (2013)
32. C.A. Pan, T.P. Ma, Appl. Phys. Lett. **37**, 714 (1980)
33. Y.-C. Lin, C.-Y.S. Chang, R.K. Ghosh, J. Li, H. Zhu, R. Addou, B. Diaconescu, T. Ohta, X. Peng, N. Lu, M.J. Kim, J.T. Robinson, R.M. Wallace, T.S. Mayer, S. Datta, L.-J. Li, J.A. Robinson, Nano Lett. **14**, 6936 (2014)
34. J.-H. Fan, P. Gao, A.M. Zhang, B.R. Zhu, H.L. Zeng, X.D. Cui, R. He, Q.M. Zhang, J. Appl. Phys. **115**, 053527 (2014)
35. X. Yu, M.S. Prévot, N. Guijarro, K. Sivula, Nat. Commun. **6**, 7596 (2015)
36. S.-C. Kung, W.E. van der Veer, F. Yang, K.C. Donavan, R.M. Penner, Nano Lett. **10**, 1481 (2010)
37. P. Wu, Y. Dai, Y. Ye, Y. Yin, L. Dai, J. Mater. Chem. **21**, 2563 (2011)
38. Y. Jiang, W.J. Zhang, J.S. Jie, X.M. Meng, X. Fan, S.-T. Lee, Adv. Funct. Mater. **17**, 1795 (2007)

39. X. Liu, L. Gu, Q. Zhang, J. Wu, Y. Long, Z. Fan, Nat. Commun. **5**, 4007 (2014)
40. Y. Guo, G. Yu, Y. Liu, Adv. Mater. **22**, 4427 (2010)
41. G. Konstantatos, E.H. Sargent, Nat. Nanotechnol. **5**, 391 (2010)
42. L. Li, L. Gu, Z. Lou, Z. Fan, G. Shen, ACS Nano **11**, 4067 (2017)
43. X. Gong, M. Tong, Y. Xia, W. Cai, J.S. Moon, Y. Cao, G. Yu, C.-L. Shieh, B. Nilsson, A.J. Heeger, Science **325**, 1665 (2009)
44. N.R. Pradhan, J. Ludwig, Z. Lu, D. Rhodes, M.M. Bishop, K. Thirunavukkuarasu, S.A. McGill, D. Smirnov, L. Balicas, A.C.S. Appl, Mater. Interfaces **7**, 12080 (2015)

# Chapter 5
# Summary and Outlook

Benifiting from the rapid development of nanomaterials growth and nano-fabrication technologies, many kinds of one- and two-dimensional materials have been widely studied in the recent years. The unique optical and physical properties of low-dimensional materials make possible high gain, fast response and wide spectrum photodetectors. In this thesis, we mainly focus on the research of photodetectors based on low-dimensional materials with novel response mechanism, including graphene near-infrared photodetector, InAs nanowire high-gain photodetector, and hybrid 2D material-nanowire photodetector. Details are as follows:

1. The graphene back-gated transistor is fabricated using mechanical exfoliation and electron beam lithography. Photocurrents produced by built-in electric field at the graphene-metal junction and by photothermoelectric effect induced by unintentional p+ doping are investigated by scanning photocurrent microscopy with near-infrared light. In addition, the amplitude of photocurrent can be tuned by changing the back-gate voltage to tune the graphene Fermi level. Finally, by deploying the graphene device as a point-like detector in a self-developed laser active imaging system, we obtain the clear infrared images of the target sample at room temperature.
2. We design core/shell-like InAs nanowires grown by chemical vapor deposition, in which a self-assembled "photogating layer" (PGL) is formed near the nanowire surface. The key function of the PGL, under light illumination, is trapping electrons generated from the core, which in return form a strong negative built-in electric field to modulate the core conductance, resulting in an anomalous photocurrent. High photoconductive gain ($\sim -10^5$) and fast response time (12 ms) are obtained in the visible spectral regime at room temperature. Photoconductive gain of $\sim -1.1$ is obtained in the near-infrared spectrum at room temperature.
3. We have demonstrated a hybrid phototransistor consisting of a single $In_2O_3$ nanowire as channel material and a multilayer $WSe_2$ nanosheet as decorating sensitizer for photodetection. With the unique design of energy band alignment of the channel and sensitizer materials, the nanowire channel can be fully depleted



N. Guo, *Infrared Photodetectors Based on Low-Dimensional Materials*,
Springer Theses, https://doi.org/10.1007/978-981-13-2838-1_5

with an external back-gate electric field, which intensifies the strength of built-in field at $WSe_2$-$In_2O_3$ interface to ensure a stronger photogating effect. Extremely low dark currents ($10^{-14}$–$10^{-13}$ A) and high responsivities ($10^4$–$10^5$ A $W^{-1}$) can be achieved simultaneously in the same operating condition. Moreover, by decoupling the light absorption from the charge transport, the phototransistors can respond to a wide spectral range from visible to near infrared. As a result, our device exhibits high detectivities up to $10^{16}$–$10^{17}$ Jones from visible to near infrared at room temperature.

For high-gain photodetectors, there are two key parameters: one is gain and the other is response time. A long lifetime of photocarriers is conducive to a high gain, but usually at the expense of response time. There is always a trade-off between the gain and the response time. Using low-dimensional materials with excellent light absorption and carrier dynamics properties to design a hybrid photodetector with novel mechanisms to solve this problem would prove to be a promising direction of future research.

FSC
www.fsc.org
MIX
Papier aus verantwortungsvollen Quellen
Paper from responsible sources
FSC® C105338

If you have any concerns about our products,
you can contact us on
**ProductSafety@springernature.com**

In case Publisher is established outside the EU,
the EU authorized representative is:
**Springer Nature Customer Service Center GmbH**
**Europaplatz 3, 69115 Heidelberg, Germany**

Printed by Libri Plureos GmbH
in Hamburg, Germany